Pitman Research Notes in Mathematics Series

Submission of proposals for consideration
Suggestions for publication, in the form of outlines and representative samples, are invited by the Editorial Board for assessment. Intending authors should approach one of the main editors or another member of the Editorial Board, citing the relevant AMS subject classifications. Alternatively, outlines may be sent directly to the publisher's offices. Refereeing is by members of the board and other mathematical authorities in the topic concerned, throughout the world.

Preparation of accepted manuscripts
On acceptance of a proposal, the publisher will supply full instructions for the preparation of manuscripts in a form suitable for direct photo-lithographic reproduction. Specially printed grid sheets are provided and a contribution is offered by the publisher towards the cost of typing. Word processor output, subject to the publisher's approval, is also acceptable.

Illustrations should be prepared by the authors, ready for direct reproduction without further improvement. The use of hand-drawn symbols should be avoided wherever possible, in order to maintain maximum clarity of the text.

The publisher will be pleased to give any guidance necessary during the preparation of a typescript, and will be happy to answer any queries.

Important note
In order to avoid later retyping, intending authors are strongly urged not to begin final preparation of a typescript before receiving the publisher's guidelines and special paper. In this way it is hoped to preserve the uniform appearance of the series.

Longman Scientific & Technical
Longman House
Burnt Mill
Harlow, Essex, UK
(tel (0279) 26721)

Titles in this series

1 Improperly posed boundary value problems
A Carasso and A P Stone
2 Lie algebras generated by finite dimensional ideals
I N Stewart
3 Bifurcation problems in nonlinear elasticity
R W Dickey
4 Partial differential equations in the complex domain
D L Colton
5 Quasilinear hyperbolic systems and waves
A Jeffrey
6 Solution of boundary value problems by the method of integral operators
D L Colton
7 Taylor expansions and catastrophes
T Poston and I N Stewart
8 Function theoretic methods in differential equations
R P Gilbert and R J Weinacht
9 Differential topology with a view to applications
D R J Chillingworth
10 Characteristic classes of foliations
H V Pittie
11 Stochastic integration and generalized martingales
A U Kussmaul
12 Zeta-functions: An introduction to algebraic geometry
A D Thomas
13 Explicit *a priori* inequalities with applications to boundary value problems
V G Sigillito
14 Nonlinear diffusion
W E Fitzgibbon III and H F Walker
15 Unsolved problems concerning lattice points
J Hammer
16 Edge-colourings of graphs
S Fiorini and R J Wilson
17 Nonlinear analysis and mechanics: Heriot-Watt Symposium Volume I
R J Knops
18 Actions of fine abelian groups
C Kosniowski
19 Closed graph theorems and webbed spaces
M De Wilde
20 Singular perturbation techniques applied to integro-differential equations
H Grabmüller
21 Retarded functional differential equations: A global point of view
S E A Mohammed
22 Multiparameter spectral theory in Hilbert space
B D Sleeman
24 Mathematical modelling techniques
R Aris
25 Singular points of smooth mappings
C G Gibson
26 Nonlinear evolution equations solvable by the spectral transform
F Calogero
27 Nonlinear analysis and mechanics: Heriot-Watt Symposium Volume II
R J Knops
28 Constructive functional analysis
D S Bridges
29 Elongational flows: Aspects of the behaviour of model elasticoviscous fluids
C J S Petrie
30 Nonlinear analysis and mechanics: Heriot-Watt Symposium Volume III
R J Knops
31 Fractional calculus and integral transforms of generalized functions
A C McBride
32 Complex manifold techniques in theoretical physics
D E Lerner and P D Sommers
33 Hilbert's third problem: scissors congruence
C-H Sah
34 Graph theory and combinatorics
R J Wilson
35 The Tricomi equation with applications to the theory of plane transonic flow
A R Manwell
36 Abstract differential equations
S D Zaidman
37 Advances in twistor theory
L P Hughston and R S Ward
38 Operator theory and functional analysis
I Erdelyi
39 Nonlinear analysis and mechanics: Heriot-Watt Symposium Volume IV
R J Knops
40 Singular systems of differential equations
S L Campbell
41 N-dimensional crystallography
R L E Schwarzenberger
42 Nonlinear partial differential equations in physical problems
D Graffi
43 Shifts and periodicity for right invertible operators
D Przeworska-Rolewicz
44 Rings with chain conditions
A W Chatters and C R Hajarnavis
45 Moduli, deformations and classifications of compact complex manifolds
D Sundararaman
46 Nonlinear problems of analysis in geometry and mechanics
M Atteia, D Bancel and I Gumowski
47 Algorithmic methods in optimal control
W A Gruver and E Sachs
48 Abstract Cauchy problems and functional differential equations
F Kappel and W Schappacher
49 Sequence spaces
W H Ruckle
50 Recent contributions to nonlinear partial differential equations
H Berestycki and H Brezis
51 Subnormal operators
J B Conway
52 Wave propagation in viscoelastic media
F Mainardi
53 Nonlinear partial differential equations and their applications: Collège de France Seminar. Volume I
H Brezis and J L Lions

54 Geometry of Coxeter groups
H Hiller
55 Cusps of Gauss mappings
T Banchoff, T Gaffney and C McCrory
56 An approach to algebraic K-theory
A J Berrick
57 Convex analysis and optimization
J-P Aubin and R B Vintner
58 Convex analysis with applications in the differentiation of convex functions
J R Giles
59 Weak and variational methods for moving boundary problems
C M Elliott and J R Ockendon
60 Nonlinear partial differential equations and their applications: Collège de France Seminar. Volume II
H Brezis and J L Lions
61 Singular systems of differential equations II
S L Campbell
62 Rates of convergence in the central limit theorem
Peter Hall
63 Solution of differential equations by means of one-parameter groups
J M Hill
64 Hankel operators on Hilbert space
S C Power
65 Schrödinger-type operators with continuous spectra
M S P Eastham and H Kalf
66 Recent applications of generalized inverses
S L Campbell
67 Riesz and Fredholm theory in Banach algebra
B A Barnes, G J Murphy, M R F Smyth and T T West
68 Evolution equations and their applications
F Kappel and W Schappacher
69 Generalized solutions of Hamilton-Jacobi equations
P L Lions
70 Nonlinear partial differential equations and their applications: Collège de France Seminar. Volume III
H Brezis and J L Lions
71 Spectral theory and wave operators for the Schrödinger equation
A M Berthier
72 Approximation of Hilbert space operators I
D A Herrero
73 Vector valued Nevanlinna Theory
H J W Ziegler
74 Instability, nonexistence and weighted energy methods in fluid dynamics and related theories
B Straughan
75 Local bifurcation and symmetry
A Vanderbauwhede
76 Clifford analysis
F Brackx, R Delanghe and F Sommen
77 Nonlinear equivalence, reduction of PDEs to ODEs and fast convergent numerical methods
E E Rosinger
78 Free boundary problems, theory and applications. Volume I
A Fasano and M Primicerio
79 Free boundary problems, theory and applications. Volume II
A Fasano and M Primicerio
80 Symplectic geometry
A Crumeyrolle and J Grifone
81 An algorithmic analysis of a communication model with retransmission of flawed messages
D M Lucantoni
82 Geometric games and their applications
W H Ruckle
83 Additive groups of rings
S Feigelstock
84 Nonlinear partial differential equations and their applications: Collège de France Seminar. Volume IV
H Brezis and J L Lions
85 Multiplicative functionals on topological algebras
T Husain
86 Hamilton-Jacobi equations in Hilbert spaces
V Barbu and G Da Prato
87 Harmonic maps with symmetry, harmonic morphisms and deformations of metrics
P Baird
88 Similarity solutions of nonlinear partial differential equations
L Dresner
89 Contributions to nonlinear partial differential equations
C Bardos, A Damlamian, J I Díaz and J Hernández
90 Banach and Hilbert spaces of vector-valued functions
J Burbea and P Masani
91 Control and observation of neutral systems
D Salamon
92 Banach bundles, Banach modules and automorphisms of C*-algebras
M J Dupré and R M Gillette
93 Nonlinear partial differential equations and their applications: Collège de France Seminar. Volume V
H Brezis and J L Lions
94 Computer algebra in applied mathematics: an introduction to MACSYMA
R H Rand
95 Advances in nonlinear waves. Volume I
L Debnath
96 FC-groups
M J Tomkinson
97 Topics in relaxation and ellipsoidal methods
M Akgül
98 Analogue of the group algebra for topological semigroups
H Dzinotyiweyi
99 Stochastic functional differential equations
S E A Mohammed
100 Optimal control of variational inequalities
V Barbu
101 Partial differential equations and dynamical systems
W E Fitzgibbon III
102 Approximation of Hilbert space operators. Volume II
C Apostol, L A Fialkow, D A Herrero and D Voiculescu
103 Nondiscrete induction and iterative processes
V Ptak and F-A Potra

104 Analytic functions – growth aspects
O P Juneja and G P Kapoor
105 Theory of Tikhonov regularization for Fredholm equations of the first kind
C W Groetsch
106 Nonlinear partial differential equations and free boundaries. Volume I
J I Díaz
107 Tight and taut immersions of manifolds
T E Cecil and P J Ryan
108 A layering method for viscous, incompressible L_p flows occupying R^n
A Douglis and E B Fabes
109 Nonlinear partial differential equations and their applications: Collège de France Seminar. Volume VI
H Brezis and J L Lions
110 Finite generalized quadrangles
S E Payne and J A Thas
111 Advances in nonlinear waves. Volume II
L Debnath
112 Topics in several complex variables
E Ramírez de Arellano and D Sundararaman
113 Differential equations, flow invariance and applications
N H Pavel
114 Geometrical combinatorics
F C Holroyd and R J Wilson
115 Generators of strongly continuous semigroups
J A van Casteren
116 Growth of algebras and Gelfand–Kirillov dimension
G R Krause and T H Lenagan
117 Theory of bases and cones
P K Kamthan and M Gupta
118 Linear groups and permutations
A R Camina and E A Whelan
119 General Wiener–Hopf factorization methods
F-O Speck
120 Free boundary problems: applications and theory, Volume III
A Bossavit, A Damlamian and M Fremond
121 Free boundary problems: applications and theory, Volume IV
A Bossavit, A Damlamian and M Fremond
122 Nonlinear partial differential equations and their applications: Collège de France Seminar. Volume VII
H Brezis and J L Lions
123 Geometric methods in operator algebras
H Araki and E G Effros
124 Infinite dimensional analysis–stochastic processes
S Albeverio
125 Ennio de Giorgi Colloquium
P Krée
126 Almost-periodic functions in abstract spaces
S Zaidman
127 Nonlinear variational problems
A Marino, L Modica, S Spagnolo and M Degiovanni
128 Second-order systems of partial differential equations in the plane
L K Hua, W Lin and C-Q Wu
129 Asymptotics of high-order ordinary differential equations
R B Paris and A D Wood
130 Stochastic differential equations
R Wu
131 Differential geometry
L A Cordero
132 Nonlinear differential equations
J K Hale and P Martinez-Amores
133 Approximation theory and applications
S P Singh
134 Near-rings and their links with groups
J D P Meldrum
135 Estimating eigenvalues with *a posteriori/a priori* inequalities
J R Kuttler and V G Sigillito
136 Regular semigroups as extensions
F J Pastijn and M Petrich
137 Representations of rank one Lie groups
D H Collingwood
138 Fractional calculus
G F Roach and A C McBride
139 Hamilton's principle in continuum mechanics
A Bedford
140 Numerical analysis
D F Griffiths and G A Watson
141 Semigroups, theory and applications. Volume I
H Brezis, M G Crandall and F Kappel
142 Distribution theorems of L-functions
D Joyner
143 Recent developments in structured continua
D De Kee and P Kaloni
144 Functional analysis and two-point differential operators
J Locker
145 Numerical methods for partial differential equations
S I Hariharan and T H Moulden
146 Completely bounded maps and dilations
V I Paulsen
147 Harmonic analysis on the Heisenberg nilpotent Lie group
W Schempp
148 Contributions to modern calculus of variations
L Cesari
149 Nonlinear parabolic equations: qualitative properties of solutions
L Boccardo and A Tesei
150 From local times to global geometry, control and physics
K D Elworthy
151 A stochastic maximum principle for optimal control of diffusions
U G Haussmann
152 Semigroups, theory and applications. Volume II
H Brezis, M G Crandall and F Kappel
153 A general theory of integration in function spaces
P Muldowney

Contributions to modern calculus of variations

Lamberto Cesari (Editor)
University of Michigan

Contributions to modern calculus of variations

Contributors:

A Bensoussan
J P Cecconi
S Cinquini
F H Clarke
I Ekeland
J Frehse
M Giaquinta
S Hildebrandt
A Marino
F Murat
C Olech
J Smoller
R Temam
C Vinti

Longman Scientific & Technical

Copublished in the United States with
John Wiley & Sons, Inc., New York

Longman Scientific & Technical
Longman Group UK Limited
Longman House, Burnt Mill, Harlow
Essex CM20 2JE, England
and Associated Companies throughout the world.

Copublished in the United States with
John Wiley & Sons, Inc., 605 Third Avenue, New York, NY 10158

First published 1987

AMS Subject Classifications: (main) 49-XX, 34-XX, 35-XX
(subsidiary) 93-XX, 90-XX

ISSN 0269-3674

British Library Cataloguing in Publication Data
Contributions to modern calculus of
variations.—(Pitman research notes in
mathematics, ISSN 0269-3674; 148)
1. Calculus of variations
I. Cesari, Lamberto
515′.64 QA315

ISBN 0-582-99458-6

Library of Congress Cataloging-in-Publication Data
Contributions to modern calculus of variations.
(Pitman research notes in mathematics,
ISSN 0269-3674; 148)
"Contributions to a symposium held in Bologna,
May 13–14, 1985 to mark the centenary of the birth of
Leonida Tonelli . . . sponsored by the University of
Bologna and by the Academy of Sciences of Bologna"—Pref.
Bibliography: p.
Includes index.
1. Calculus of variations—Congresses. 2. Tonelli,
Leonida, 1885– . I. Cesari, Lamberto.
II. Università di Bologna. III. Accademia delle
scienze dell'Istituto di Bologna. IV. Series.
QA315.C65 1986 515′.64 86-15700
ISBN 0-470-20378-1 (USA only)

Printed and bound in Great Britain by
Biddles Ltd, Guildford and King's Lynn

Contents

Preface

The papers collected in this volume are contributions to a symposium, held in Bologna, May 13-14, 1985 to mark the centenary of the birth of Leonida Tonelli. The symposium was sponsored by the University of Bologna and by the Academy of Sciences of Bologna. Another celebration had taken place a few days earlier at the National Academy of Science in Rome, the Lincei, to honour both Leonida Tonelli and Mauro Picone.

Some of the papers from the conference in Bologna have been translated from Italian. One of the papers was contributed by a colleague, who was unable to attend the conference.

Particular thanks are due to Dario Graffi, President of the Academy of Science of Bologna, and to Ilio Galligani, Director of the Department of Mathematics of the University of Bologna, for their warm personal interest; without their support, the conference could not have taken place. I wish to express my personal thanks to Professor Gaetano Fichera for having agreed to help our Conference as a member of the Organizing Committee after having cared for the Conference in Rome under the frame of the Academy of Lincei. And many thanks are also due to the members of the Organizing Committee (in alphabetical order) Professor Vinicio Boffi (Treasurer), Professor Vittorio Bononcini (Secretary), Professor Sandro Graffi, and Professor Angelo Pescarini for their personal commitment and unfailing support.

We are indebted to the staff of Longman Scientific and Technical Division for their expertise in the production of this book, and to Terri Moss for having efficiently and competently typed the entire volume.

Lamberto Cesari

List of contributors

A. Bensoussan:	Université de Paris-Dauphine and INRIA, Paris, France.
J.P. Cecconi:	Istituto di Matematica Applicata, Università di Genova, 16132 Genova, Italy.
S. Cinquini:	Dipartimento di Matematica, Università di Pavia, 27100 Pavia, Italy.
F.H. Clarke:	Centre de Recherches Mathématiques, Université de Montréal, Montréal, Quebec, Canada H3C 3J7.
I. Ekeland:	Centre de Recherche de Mathématiques de la Decision, Université de Paris IX, Paris, France.
J. Frehse:	Institut Angewandte Mathematik, Universität Bonn, 5300 Bonn, West Germany.
M. Giaquinta:	Istituto di Matematica Applicata, Facoltà Ingegneria, 50134 Firenze, Italy.
S. Hildebrandt:	Mathematisches Institut, Universität Bonn, 5300 Bonn, West Germany.
A. Marino:	Dipartimento di Matematica, Università di Pisa, 56100 Pisa, Italy.
F. Murat:	Laboratoire d'Analyse Numérique, Université de Paris VI, 75320 Paris, France.
C. Olech:	Institute of Mathematics, Polish Academy of Science, Warsaw, Poland.
J.A. Smoller:	Department of Mathematics, University of Michigan, Ann Arbor, Michigan 48109, USA.
R. Temam:	Laboratoire d'Analyse Numérique, Université de Paris XI, 91405 Orsay, France.
C. Vinti:	Dipartimento di Matematica, Università degli Studi, 06100 Perugia, Italy.

L CESARI

A tribute to Leonida Tonelli: Leonida Tonelli (1885–1946) and his 20th century legacy

Leonida Tonelli, born in Gallipoli on the 19th of April 1885 from parents of the venetian region, studied at the University of Bologna (1902-07), and taught there until 1914. It was in Bologna that he had the illuminating influence of C. Arzelà and S. Pincherle, both interested in functional analysis from different view points. In 1884 Vito Volterra had proposed the concept of functional in the real field, and he considered the integrals of the calculus of variations as typical examples of functionals. The ideas of Vito Volterra were actively cultivated by Arzelà in Bologna.

We know that in the 19th century it was thought that any non-negative integral of the calculus of variations had an absolute minimum, and indeed Riemann, from such a belief concerning the Dirichlet integral, had derived important results in analysis, such as the existence of harmonic functions in a bounded domain having prescribed values on the boundary. It was Weierstrass who criticized Riemann's argument, and this and other difficulties gave rise to the discussion on the foundations of Analysis in which Arzelà, Ascoli, Dini, Schwartz, Peano, Hilbert took lively parts. It was Arzelà in Bologna who tried unsuccessfully to obtain a direct proof of the existence of the minimum for the Dirichlet integral. Hilbert in 1900 obtained such a direct proof, but his proof relied on specific properties of harmonic functions. These ideas were followed with enthusiasm by Beppo Levi (1907), by H. Lebesgue, and G. Fubini.

Meanwhile R. Baire (1905) had introduced his concept of semicontinuity for real valued functions showing that, in a compact domain, lower semicontinuous functions had an absolute minimum, and upper semicontinuous functions had an absolute maximum.

Tonelli became acutely aware that a theory was needed for the direct existence proof of maxima and minima of general problems of the calculus of variations. It was in Bologna that Tonelli realized that in the Ascoli-Arzelà compactness theorems and Baire semicontinuity transferred from real functions to the functionals of the calculus of variations, he had the perfect tools for a "direct method in the calculus of variations based on semicontinuity" for an existence theory of maxima and minima. By this it was meant, as it is meant today, that to prove the existence of the absolute minimum of a functional I(u) in a class K, one has to prove first that the functional has a finite infimum for u in K, that the functional is lower semicontinuous with respect to some type of convergence, and finally that there is some minimizing sequence which converges in the stated type of convergence toward an element u_0 which belongs to K, and which then realizes the absolute minimum of I(u) in K. The type of convergence is not assigned, but at the time, and for many more years, the natural convergence was the uniform convergence of the trajectories.

Indeed, when very young, Tonelli proved in 1911 the existence of the absolute minimum for general positive regular integrals of the calculus of variations over parametric continuous rectifiable curves in $\mathbb{R}^2$. The regularity was only a strong form of convexity, actually a strong form of the Legendre and Weierstrass conditions, and this assumption implies the lower semicontinuity of the integrals with respect to uniform convergence. The realization was reached that the classical Legendre and Weierstrass conditions are not so much conditions for a minimum or a maximum, but conditions for the lower or upper semicontinuity of the integral. For parametric integrals the conditions, required by Tonelli at the time, are expressed today by requesting only continuity and convexity.

In 1914-15 Tonelli proved the existence of the absolute minimum for ordinary curves in $\mathbb{R}^2$, (i.e. non-parametric), namely, absolute continuous functions (AC), and under conditions at the time, which are expressed today by requesting only continuity, convexity, and coercivity. Indeed, these conditions imply the properties of lower semicontinuity and compactness for the minimizing sequence which are essential for the conclusion.

Tonelli left Bologna for a chair in Cagliari and then Parma (1914), and from here he volunteered for active duty on the front in World War I. Afterwards, he collected the new ideas and the impressive amount of work he had accumulated in so few years in a complete theory, which appeared as his main opus, the "Fondamenti di Calcolo delle Variazioni", a complex of 1150 pages, two volumes, in 1921-23.

It was in 1922 when the University of Bologna called him to a chair, which he held until 1930, when he accepted a move to the University of Pisa. In the years in Bologna and then in Pisa he set the course and influenced a great many people, to name a few, A. Mambriani, E. Lindner, D. Graffi, and S. Cinquini in Bologna, and B. Manià, A. Del Chiaro, G. Stampacchia, E. Magenes, E. Baiada, S. Faedo, L. Amerio, and myself in Pisa. Later, in Pisa, he influenced L. Giuliano, J. Cecconi, G. Torrigiani, R. Cisbani, C. Santacroce, G. Ottaviani, A. Pedrini, B. Colombo, and M. Dolcher. In the twenties, in Bologna, Tonelli married Maria Rondelli, a woman of great spirit and courage, who was going to be his life companion.

The Tonelli basic results on rectifiable continuous curves (1908-12) had prepared the way for the first Tonelli existence theorems, and now a spate of important results filled a new area so created in the calculus of variations. In the twenties Tonelli presented his ideas in international gatherings (Toronto 1924, Moscow 1925, Bologna 1928). Mathematicians such as H. Hahn, C. Carathéodory, M. Nagumo contributed to the new ideas. From a 1922 remark of Tonelli, J. Lavrentiev (1925) presented an example showing that the infimum of a functional in the AC class of functions may be actually lower than the infimum of the same functional in the more restricted class of all Lipschitz functions, or the traditional class C^1. This is the Lavrentiev phenomenon to which Tonelli, B. Manià, S. Cinquini further contributed. Quite recently T.S. Angell unified their result by the use of a concept, property D, which had been introduced by Cesari and Suryanarayana in questions concerning the existence of optimal solutions. In 1913 Tonelli obtained an existence theorem for isoperimetric problems in $\mathbb{R}^2$, and he later improved his result several times. In the same year he gave a rigorous proof of the property of minimum of the sphere.

In the early thirties in Pisa, Manià, Cinquini (who had followed Tonelli from Bologna), and A. Del Chiaro obtained further results in the new development in the calculus of variations. Again in the early thirties, L.M. Graves and E.J. McShane of the School of Chicago contributed essentially to the new ideas. Particularly McShane contributed to the existence theory for non-parametric non-coercive integrals, showing that they can be reduced to parametric problems, albeit possibly singular, and McShane overcame great difficulties in such a program which became an important addition to the theory of Tonelli. Later Cinquini and Faedo obtained existence theorems for problems of the calculus of variations on infinite intervals, problems of interest in econometrics. Quite recently, after the theory of optimal controls became a part of the calculus of variations, R.F. Baum and also D.A. Carlson deepened the existence theory for such problems. In the early thirties it was already clear that Tonelli had created a drastic change in attitude and understanding in the calculus of variations, which by now had become a chapter of functional analysis, namely the chapter dealing with minima and maxima of functionals. In 1936 Volterra and Pérès asked Tonelli to write such a chapter for their book on functional analysis, and while Tonelli is mostly known for the new process leading to existence theorems, he contributed in essential ways also to problems concerning the regularity of the optimal solutions and to the necessary conditions. Indeed, he showed that the optimal solutions, whose existence was proved by his theorems, had enough differentiability properties to satisfy the Euler equation everywhere, except possibly at the points of a closed exceptional set of measure zero, while the integrability of the two sides of the Euler equation had to be guaranteed by independent assumptions. At the time it was not known if the exceptional closed set of measure zero could occur, or whether it could actually happen that the two members of the Euler equation could not be integrable. Quite recently, Ball-Mizel and Clarke-Vinter showed by examples that both phenomena may actually occur.

Besides his fundamental work in the calculus of variations, Tonelli made important contributions to real analysis, trigonometrical series, applied mathematics, and partial differential equations theory.

Meanwhile, Tonelli had obtained his basic results on rectifiable continuous non-parametric surfaces (1926) where he introduced his globally continuous functions of $k \geq 1$ variables of bounded variations (BVT) and absolutely continuous (ACT), and in 1928 he obtained his first existence theorems for general multiple integrals of the calculus of variations in classes of globally continuous ACT functions. Cinquini, Manià, and Del Chiaro in Pisa, and McShane and C.B. Morrey in the US contributed to this new development of Tonelli's theory. In particular, McShane and Morrey also characterized classes of parametric surfaces of finite Lebesgue area admitting representations in terms of globally continuous ACT functions (monotone surfaces, saddle surfaces). The new ideas also influenced J. Douglas and T. Radó, who independently from each other, proved (in 1930) the existence of a solution to the Plateau problem, that is, the problem concerning the existence of a continuous parametric surface of the type of the disc spanning a given continuous closed simple curve in $\mathbb{R}^3$. More elegant proofs for the same Plateau problem were soon obtained by McShane (1933), Tonelli (1936), and R. Courant, while Douglas considered the same problem in classes of surfaces of given finite topological type. W.H. Fleming then showed examples where the actual minimum of the area is achieved by surfaces of infinite topological type. A rich literature followed on the minimum of the area in classes of surfaces of undetermined type (W.H. Fleming, E.R. Reifenberg, H. Federer), and these studies are now included in F.T. Almgren's (1965) theory of varifolds and subsequent work of R. Osserman and E. Bombieri (1985).

In the late forties the theory of Tonelli was enriched by the use of absolute continuous functions of $k \geqq 1$ independent variables in a domain G of $\mathbb{R}^k$, which are not necessarily globally continuous. Already in the early years of this century Beppo Levi, H. Lebesgue, G. Fubini had considered classes of such functions ($W^{1,1}(G)$, $W^{1,2}(G)$ in present day's notation) for the study of multiple integrals of the calculus of variations. So did G.C. Evans in 1920 and N. Nikodym in 1933 in other parts of analysis. It was natural that C.B. Morrey (1940, 1943), who had worked so deeply in Tonell's theory, realized that abandoning the global continuity requirement in the ACT concept could lead to more general and satisfactory existence theorems for multiple integrals. Meanwhile A.G. Sobolev, for exigences deeply felt in the theory of partial differential equations, had introduced his classes of functions, or spaces $W^{m,p}(G)$ (1938). In the years between World War II and 1950 Stampacchia also matured ideas similar to those of Morrey with analogous results.

On the other hand, the new existence theorems prove only that the optimal solutions are elements of suitable classes $W^{m,p}(G)$, and therefore it is not known even if they are continuous in situations where one expects them to be smooth, even analytic. Morrey, in 1940, and then more deeply in 1943, proved that the optimal solutions have Hölder first derivatives in the interior of the domain. Stampacchia also concerned himself with the question of the regularity of the solutions (1952).

These new ideas generated a tremendous amount of research, particularly in the study of partial differential equations. Let us mention here only the work of Sobolev himself, and of V.I. Kondrachov (1945), A.G. Sigalov (1952),

F. John (1953), J. Deny and J.L. Lions (1955), F. Gagliardo (1958), S.F. Uspenskii (1960), J.L. Lions (1961), J.L. Lions and E. Magenes (1961), W.H. Fleming (1962).

After the work of Morrey and Stampacchia, most of the research concentrated on the proof of the differentiability and analyticity of the optimal solutions for multiple integrals. A great number of mathematicians took part in this program. I mention here Morrey himself, E. DeGiorgi, S. Nash, G. Stampacchia, L. Nirenberg, and more recently M. Giaquinta, E. Giusti, M. Marino, M. Miranda, F. Ferro, and B. Scheurer.

Problems of the calculus of variations were considered with constraints on the state variable (obstacles), and a great deal of research was dedicated to the proof of the regularity of the optimal solutions near the obstacle. I mention here only the work of H. Lewy, G. Stampacchia, M. Miranda, S. Hildebrandt.

Also, G. Fichera (1963), G. Stampacchia (1964), and J.L. Lions developed the theory of "variational inequalities", a theory which includes calculus of variations proper, has the most relevant applications, and has generated a great amount of work. A problem of elasticity of this kind was first proposed by A. Signorini (1959) and studied by the direct method by G. Fichera (1963).

The theory of Tonelli had a further enlargement through the work of Morrey, who studied multiple integrals of the calculus of variations for n functions of k independent variables in a domain G, that is, for a mapping $T : G \subset \mathbb{R}^k \to \mathbb{R}^n$, $1 < k \leq n$, sufficiently smooth. Morrey noted that, for these mappings, the convexity of the integrand with respect to the gradient, which is a kn-vector, is neither a necessary nor a sufficient condition for lower semicontinuity. He then defined a concept of quasi convexity for the integrand which is a necessary and in a sense sufficient condition for lower semicontinuity. This concept is difficult to handle, and Morrey indicated a slightly more restrictive class of integrands, the polyconvex ones, those which can be written as convex functions of all $\binom{n+k}{k} - 1$ Jacobians of all orders from one to k (including the nk first order partial derivatives themselves). Work was done on polyconvex integrals by Morrey and by E. Silverman. A new development on these integrals was initiated by J. Ball in 1977 who discovered that in nonlinear elasticity the integrals representing the total energy of the system are always polyconvex. J. Ball then obtained existence theorems for the minimum of such integrals in terms of Sobolev spaces. Much more work followed by Ball himself and N. Fusco. Recently, L. Tartar and

F. Murat have applied their theory of "compactness by compensation" to polyconvex integrals.

Another extension of Tonelli's theory occurred in the years 1939-52 through surface area theory. After the preparatory work of C.B. Morrey, E.J. McShane, T. Radó, and R. Caccioppoli, Cesari resumed the study of parametric continuous mappings $T : G \subset \mathbb{R}^k \to \mathbb{R}^n$, or (k,n)-maps, under the sole assumption that T is parametric and continuous. Relevant here are the $\binom{n}{k}$ (flat) maps T_α, $\alpha = 1,\ldots,\binom{n}{k}$, which are the projections of T on the coordinate k-dimensional subspaces of $\mathbb{R}^n$. In terms of topological degrees, it is possible to define total variations $V(T_\alpha)$ and absolute continuity (AC) of the flat maps T_α, and also a lower area A(T) and a Peano area P(T) of T, besides the Lebesgue area L(T). Moreover, in terms of Radon-Nikodym derivatives of suitable set functions, it is possible to define generalized Jacobians $J_\alpha(u)$, $u \in G$ (a.e.), of the flat maps T_α. For k = 2, Cesari proved (1941-42) that the relations between L(T) and the total variations $V(T_\alpha)$, between L(T) and the generalized Jacobians, are very similar to those for length of curves, and for non-parametric surfaces proved by Tonelli. Moreover, Cesari proved that A(T) = L(T) for all continuous maps T, $G \subset \mathbb{R}^2 \to \mathbb{R}^n$, and J. Cecconi proved that L(T) = P(T). Again, for k = 2, Cesari introduced a concept of the parametric integral I(T) of the calculus of variations defined solely in terms of continuity and finiteness of the Lebesgue area of T, and proved necessary conditions and sufficient conditions for the lower semicontinuity of I(T), and an existence theorem for the Plateau problem for positive convex parametric integrals (1952). Thus, a far reaching and deep theory of continuous (k,n)-maps, mostly for k = 2, was developed, with essential contributions of L.C. Young, T. Radó, J.W.T. Youngs, V. Bononcini, J. Cecconi, P.V. Reichelderfer, E.J. Mickle, H. Federer, E.J. Neugebauer, E.R. Fullerton.

In 1936 the direct method in the calculus of variations based on semicontinuity was further enriched by a remarkable enlargement of the class of curves (or surfaces) among which to search for an optimal one: the "generalized curves" of L.C. Young. The most diverse interpretations lie at the foundation of this concept, from taking the convex hull of certain sets, to a measure and probabilistic interpretation, or to a linear functional analysis approach, particularly with the aim of guaranteeing optimal solutions even to problems with non-convex integrands. A great deal of work was done in this direction, first by L.C. Young himself and W.H. Fleming. Later, after the theory of optimal control became part of the calculus of variations, R.V. Gamkrelidze (1964) observed further wide applications of this new approach, even in proofs of necessary conditions. Further enlargements of the concept of generalized solutions were then proposed and systematically developed by I. Ekeland, R. Temam, and J. Warga.

Let us mention here that F.H. Clarke has developed a theory of necessary conditions in the calculus of variations for non-smooth integrands, and in this endeavour he has deepened some remarks made by Tonelli in 1915. Also, R.T. Rockafellar has developed a duality theory in convex analysis with remarkable applications to the calculus of variations, and F.H. Clarke and I. Ekeland have realized a duality approach to the concept of action of theoretical mechanics.

In 1936, before Morrey and before Sobolev, Cesari proposed to abandon global continuity in Tonelli's concept of bounded variation. This had to be done with care, in particular by systematically disregarding the values taken by the function in suitable sets of measure zero. The resulting concept, say of BV functions, for short, in a bounded domain G of $\mathbb{R}^k$, was shown by Cesari to be invariant with respect to 1-1 Lipschitzian transformations in $\mathbb{R}^k$, and to have a bearing in the theory of almost everywhere convergence of multiple Fourier series. In 1950 F. Cafiero proved relevant properties of compactness in $L_1(G)$ of such functions, and in 1968 E. Conway and J.A. Smoller proved the existence of weak locally BV solutions (shock waves) to the Cauchy problem for nonlinear conservation laws as hyperbolic partial differential equations. In 1954, after the theory of distributions became known, K. Krickeberg proved that the BV functions in G are exactly those $L_1(G)$-functions whose first order distributional partial derivatives are finite measures. Thus, a second equivalent definition of BV functions ensued, and much work followed in the line of the new definition. I mention here only W.H. Fleming, A.I. Volpert, M. Giaquinta, G. Anzellotti, E. De Giorgi, E. Gagliardo, E. Giusti, G. Da Prato, M. Miranda. Existence theorems for (discontinuous) BV solutions of non-coercive positive problems of the calculus of variations for single and multiple integrals have been proved by O. Caligaris, P. Oliva, F. Ferro, and more recently by Cesari, P. Brandi and A. Salvadori. Also, recently, R. Temam and G. Strand have discussed problems of plasticity in terms of multiple integrals over discontinuous functions of "bounded deformation", also defined in terms of distributions.

In the years 1955-59, L.S. Pontryagin, possibly motivated by problems of space mechanics, proposed a new form of problems of optimization in which certain control functions, taking their values in given "control sets", generate, together with the boundary data, certain solutions, or trajectories, of the mechanical systems, usually monitored by ordinary differential equations or systems of such equations. It is expected that the control functions may take their values even on the boundary of the control sets, where optimal solutions often lie. The theory came at the right moment of the technical development, and had an immense response in the applied field as well as in theoretical studies. The book by L.S. Pontryagin, R.V. Gamkrelidze, V.G. Boltyanskii, and E.F. Mishchenko (1962) helped to spread the theory, which was first used in problems of space mechanics, later in important problems of econometrics, and then in problems of architectural structures. Pontryagin had already conceived of the trajectories as being absolutely continuous (AC) (and the control functions as measurable), and by doing so he adhered closely to the Tonelli approach. Perhaps, the most often mentioned result is Pontryagin's maximum principle, a wide ranging necessary condition for optimal solutions, and in the original proof of it, use was made of certain 1939 results of McShane's for classical Lagrange problems. A great many authors presented new proofs of the principle. I just mention here Pontryagin, Gamkrelidze, L. Neustadt, H. Halkin, M.R. Hestenes, E.J. McShane, J. Warga, Cesari and R.F. Baum, R. Conti, P. LaSalle, H. Hermes. A parallel approach to optimal control theory is R. Bellman's dynamic programming theory, which also generated an immense amount of work, and whose deep relations with optimal control have been successfully investigated by Boltyanskii.

The first and direct existence theorem for problems of optimal control with compact control sets was obtained by A.F. Filippov (1962), and E.J. McShane and R.B. Warfield in 1967 contributed to a reassessment of his proof. Also in 1967, by a reinterpretation of the process, Cesari showed that the concept of lower semicontinuity was still at the base of Filippov's proof, and showed that, under standard conditions of continuity, convexity, and coerciveness, also existence theorems of the Tonelli type could be proved even for unbounded control sets. Actually, it was clear that problems of optimal control, by a process of deparametrization, could be reduced to problems of the calculus of variations with constraints on the derivatives, and that usual lower semicontinuity theorems could be equivalently expressed in terms of closure and lower closure theorems, as Cesari showed, but still in the frame of Tonelli's direct method based on lower semicontinuity.

Meanwhile, R. Temam, A.D. Ioffe and other mathematicians pointed out that the concept of convergence represented by "uniform convergence of the trajectories and weak convergence in L_1 of their derivatives" had great advantages over the mere "uniform convergence of the trajectories" used, e.g. by Tonelli and McShane. This remark, worded for single integrals, has a natural parallel in terms of multiple integrals and systematic use of strong and weak convergences in suitable Sobolev spaces. By the work of a great many mathematicians, a rich existence theory, based on semicontinuity, ensued from these remarks.

Cesari, with the help of a number of collaborators such as L.H. Turner, M.B. Suryanarayana, T.S. Angell, F.H. Baum, P.J. Kaiser, J.R. LaPalm, T. Nishiura, D.A. Sanchez and P. Pucci, proved a great many existence theorems, for problems of the calculus of variations and of optimal control theory, some based on convexity and coercivity à la Tonelli, some in which the coercivity is relaxed along certain systems of lines, or at the points of an exceptional "slender" set in the terminology of Turner, some for integrals of slow growth in the manner of McShane, some based on controlled growth, unifying views of E.H. Rothe, D. Berkovitz, and F. Browder, some based on Lyapunov's theorem on the convexity of the range of any vector valued non-atomic measure.

Tonelli in his initial work in 1911 on integrals of the calculus of variations on rectifiable parametric curves, defined integral in the sense of Weierstrass, as being invariant with respect to any change in the representation of the curve, and therefore representing a deep seated property of the curve. Later, he abandoned the use of Weierstrass integrals for Lebesgue integrals and sole AC representation of the curves. Tonelli later returned to questions of limit and integration, and relevant work was subsequently done by S. Faedo, E. Baiada and C. Vinti in this direction. Also, in 1925, Tonelli proposed a very elementary approach to Lebesgue integration for real functions of one or more independent variables, by-passing measurability, namely, for the class of quasi-continuous functions as equivalent to the class of the measurable ones.

However, there are situations where the use of Weierstrass type integration is unavoidable. For instance, in the theory of integration over (k,n)-maps, with the sole assumptions of continuity and finiteness of the area, Cesari (1948-51) used a Weierstrass type integral based on a process of limit over certain set functions which had been defined in terms of topological degrees for the $\binom{n}{k}$ corresponding flat maps T_α. Later, in 1962, Cesari extended this process of limit and integration to a general setting, and for set functions assumed to be only quasi-additive. In this situation Cesari defined an abstract Burkill type integral, and in turn a Weierstrass type integral over a continuous variety. Under hypotheses, the quasi-additive set functions define regular measures and the Weierstrass integral has a representation as a Lebesgue-Stieltjes integral. Many mathematicians developed these ideas, showing in particular, that a number of processes of limit and integration can be framed in terms of quasi-additive set functions. I mention here only the work of A.W.T. Stoddart, G. Warner in the US, and C. Vinti, M. Boni, C. Gori, P. Brandi, A. Salvadori, A. Averna, A. Fiacca, C. Lodovici, P. Pucci in Italy, on simple and multiple integrals of the calculus of variations, and their properties and representation. I can only refer here to recent expositions of C. Vinti.

The present exposition is contained in a more detailed one with extensive bibliography: L. Cesari, L'opera di Leonida Tonelli e la sua influenza nel pensiero scientifico del secolo. This more detailed exposition is appearing in the volume dedicated to Mauro Picone and Leonida Tonelli, Accademia Nazionale dei Lincei, Roma 1986. Most of the notes and memoirs of Tonelli are collected in the four volume series L. Tonelli, Opere Scelte, Cremonese 1960-63.

We have only mentioned above a few of the many and vast ramifications of Tonelli's theory. He has not had the satisfaction of seeing the immense developments of his own ideas. However, by the late thirties, Tonelli must have seen clearly the success of his lifetime's work. The hardships of World War II were the chief cause of the recrudescense of a lingering disease. Leonida Tonelli died suddenly on the 12th of March 1946 in Pisa where he was given national recognition.

L. Cesari
Department of Mathematics
University of Michigan
Ann Arbor
Michigan 48109
USA

A BENSOUSSAN AND J FREHSE

On Bellman equations of ergodic type with quadratic growth Hamiltonians

INTRODUCTION

Consider the following stochastic dynamic system (written slightly formally to avoid details)

$$dx = v(t)dt + \sqrt{2}\, dw \qquad x(0) = z$$

where v(t) is the control, and w(t) is a standard Wiener process. We want to minimize the cost function

$$J_{\alpha,x}(v(\cdot)) = E\int_0^\infty e^{-\alpha t}[\ell(x(t)) + \frac{1}{4}\, v(t)^2\,]dt$$

where ℓ is a periodic function (period 1 in all components). We assume full observation. Define the value function

$$u_\alpha(x) = \min_{v(\cdot)} J_{\alpha,x}(v(\cdot))$$

then u_α is solution of the Bellman equation

$$-\Delta u_\alpha + (Du_\alpha)^2 + \alpha u_\alpha = \ell$$

and the solution is periodic, since ℓ is periodic. The ergodic control problem deals with the case $\alpha \to 0$.

The objective of this paper is to study more general Bellman equations of the type

$$(1) \qquad Au_\alpha + \alpha u_\alpha = H(z,Du_\alpha)$$

with quadratic growth Hamiltonians. The case of linear growth Hamiltonians has been already studied by A. Bensoussan [1], whereas more general growth conditions have been considered by Gimberg [2]. However we do not make the regularity assumptions of Gimberg in this paper, and our methods are different. The results are of the following type. There exists a pair (z,ρ) where z is a periodic function and ρ is a scalar, satisfying

(2) $-\Delta z + \rho = H(x,Dz)$.

The basic point is the availability of an H^1-estimate for u_α. This suffices to prove the existence of a regular solution to (2). The fact that the H^1-estimate for the approximating problem is sufficient to derive an L^∞ bound is of interest in itself. Indeed the class of equations we consider admits also unbounded H^1-solutions,

$$\text{e.g.} \quad \Delta u = (n-2)|Du|^2, \quad u = \log|x|, \; n \geq 3.$$

Acknowledgements

The first author would like to thank the Von Humboldt Stiftung, for supporting his stay at Bonn University.

1. SETTING THE PROBLEM

1.1 Assumptions and Notations

Consider the 2nd order differential operator

$$A = -\sum_{i,j} \frac{\partial}{\partial x_i} a_{ij} \frac{\partial}{\partial z_j} + \sum_i a_i \frac{\partial}{\partial x_i} \tag{1.1}$$

where

$$a_{ij}(x) \text{ periodic} \tag{1.2}$$

with period 1 in all components ($x \in \mathbb{R}^n$) (This assumption is made to simplify the notation, but is by no means necessary). We take $Y =]0,1[^n)$ and assume

$$\begin{aligned} &\sum_{ij} a_{ij}(x)\xi_i\xi_j \geq \beta|\xi|^2, \quad \forall \xi \in \mathbb{R}^n, \quad \beta > 0, \\ &\qquad a_{ij} \text{ periodic bounded.} \end{aligned} \tag{1.3}$$

We consider now a function $H(x,p)$ on $R^n \times R^n$, such that

$$\begin{aligned} &H(x,p) \text{ is periodic in } x, \text{ Borel}, \\ &-k_1(1 + |p|^2) \leq H(x,p) \leq k_0(1 - |p|^2), \\ &\qquad \left|\frac{\partial H}{\partial p}\right| \leq k_2(1 + |p|). \end{aligned}$$

A typical function H will be a Hamiltonian arising in Bellmann equations, in which

$$H(x,p) = \inf_v \{\ell(x,v) + p,g(x,v)\} \tag{1.5}$$

where

$$\ell(x,v) \text{ is Borel, periodic in } x, \tag{1.6}$$

$$\ell_0 I \leqq \ell_{vv}(x,v) \leqq \ell_1 I,\ \ell_0,\ell_1 > 0,$$

$$I = \text{identity matrix},$$

$$g(x,v) = g_0(x) + g_1(x)v, \tag{1.7}$$

where g_0, g_1 are Borel, periodic, bounded, $g_1 g_1^*$ is invertible.

Let $\hat{V}(x,p)$ be the pair of minimum in (1.5). It is uniquely defined by the equation

$$\ell_v(x,\hat{V}) + g_1^*(x)p = 0. \tag{1.8}$$

But

$$\ell_v(x,\hat{V}) = \ell_v(x,0) + \left[\int_0^1 \ell_{vv}(x,\lambda\hat{V})d\lambda\right]\hat{V},$$

hence we can write

$$\hat{V} = -\left[\int_0^1 \ell_{vv}(x,\lambda\hat{V})d\lambda\right]^{-1} (\ell_v(x,0) + g_1^*(x)p). \tag{1.9}$$

On the other hand

$$\begin{aligned} H(x,p) &= \ell(x,\hat{V}) + p\cdot(g_0 + g_1\hat{V}) \\ &= \ell(x,0) + p\cdot g_0 - \int_0^1\int_0^1 \lambda\,\ell_{vv}(x,\hat{V}(1-\lambda\mu))\hat{V}^2 d\lambda d\mu. \end{aligned} \tag{1.10}$$

Write

$$P = \int_0^1\int_0^1 \lambda\,\ell_{vv}(x,\ \hat{V}(1-\lambda\mu))d\lambda d\mu,$$

$$Q = \left[\int_0^1 \ell_{vv}(x,\lambda\hat{V})d\lambda\right]^{-1}.$$

We deduce

$$H(x,p) = \ell(x,0) + p\cdot g_0 - PQ(\ell_v(x,0) + g_1^* p)\cdot Q(\ell_v(x,0) + g_1^* p) \qquad (1.11)$$

and from the assumption (1.6), (1.7) it follows at once that

$$\gamma_0 I \leqq g_1 QPQ g_1^* \leqq \gamma_1 I,$$
$$\gamma_0, \gamma_1 > 0, \text{ constant}; \qquad (1.12)$$

hence the assumption (1.4) is satisfied, since also

$$\frac{\partial H}{\partial p} = g_0(x) + g_1(x)\hat{V}.$$

1.2 The problem

Consider the Bellman equation

$$Au_\alpha + \alpha u_\alpha = H(x, Du_\alpha),$$
$$u_\alpha \text{ periodic.} \qquad (1.13)$$

The problem (1.3) has for $\alpha > 0$, one and only one solution in $W^{2,p}(Y)$, $\forall\, p, 2 \leqq p < \infty$. We are interested in the behaviour of u_α as $\alpha \to 0$.

1.3 Statement of the result

Our objective is to prove the following

Theorem 1.1 *Under the assumptions* (1.2), (1.3), (1.4) *one has*

$$\alpha u_\alpha(x) \to \rho,$$
$$u_\alpha(x) - u_\alpha(x_0) \to z \text{ in } W^{2,p}(Y) \text{ weakly}, \qquad (1.14)$$

where x_0 *is arbitrary in* Y, ρ *is a constant and the pair* z, ρ *is a solution of*

$$Az + \rho = H(x, Dz),$$
$$z \text{ periodic.} \qquad (1.15)$$

The constant ρ *is uniquely defined, the function* z *is uniquely defined if one adds the condition* $z(x_0) = 0$.

2. PROOF OF THEOREM 1.1

2.1 Preliminaries

Lemma 2.1 *One has the result*

$$|\alpha u_\alpha(x)| \leqq C. \tag{2.1}$$

Proof This follows from maximum principle considerations. Formally, assuming additional regularity so that the solution of (1.9) becomes classical, let x_α be the point of maximum of u_α (in Y), then $Du_\alpha(x_\alpha) = 0$, and $Au_\alpha(x_\alpha) > 0$. Therefore from the equation (1.9), it follows that

$$\alpha u_\alpha(x_\alpha) < H(x_\alpha, 0) \leqq k_0,$$

hence

$$\alpha u_\alpha(x) < k_0.$$

Similarly, at a point of minimum $\bar{x}_\alpha$, one has $Du_\alpha(\bar{x}_\alpha) = 0$, $Au_\alpha(\bar{x}_\alpha) < 0$, hence

$$\alpha u_\alpha(\bar{x}_\alpha) > H(\bar{x}_\alpha, 0) \geqq -k_1,$$

hence

$$\alpha u_\alpha(x) \geqq -k_1$$

and we obtain the desired result. If the solution is not classical, a different but standard argument is needed, but (2.1) is still true with

$$C = \max(k_0, k_1).$$

In the sequel we shall consider some particular point x_α of Y. Define $G^\alpha(x)$ to be the Green function associated to x_α, i.e., defined by

$$\sum_{i,j} \int_Q a_{ji} \frac{\partial G^\alpha}{\partial x_j} \frac{\partial \phi}{\partial x_i} \, dx + \sum_i \int_Q a_i \cdot G^\alpha \frac{\partial \phi}{\partial x_i} \, dx = \phi(x_\alpha) \tag{2.2}$$

$\forall\ \phi$ with compact support in some domain $Q \supset \hat{Y}$ and $\phi \in H^1(Q)$; Q is chosen independently of α. The function G^α satisfies the estimate (c.f.e.g. [3], [4])

$$\frac{\alpha_0}{|x-x_\alpha|^{n-2}} \leqq G^\alpha(x) \leqq \frac{\alpha_1}{|x-x_\alpha|^{n-2}} \tag{2.3}$$

$\alpha_0, \alpha_1 > 0$ independent of α, if $n \geqq 3$;

$$\alpha_0 \log \frac{1}{|x|} \leqq G^\alpha(x) \leqq \alpha_1 \log \frac{1}{|x|} \tag{2.4}$$

if $n = 2$.

In the sequel we assume $n \geqq 3$, to fix the ideas.

Lemma 2.2 *One has the estimate*

$$\int_Y |Du_\alpha|^2 \, dx \leqq C. \tag{2.5}$$

Proof Integrate (1.13) over Y we get

$$\sum_i \int_Y a_i \frac{\partial u_\alpha}{\partial x_i} dx + \alpha \int_Y u_\alpha dx = \int_Y H(x, Du_\alpha) dx$$

and from (1.4)

$$\leqq k_0 - k_0 \int_Y (Du_\alpha)^2 \, dx$$

from which one easily deduces (2.5).

2.2 Main estimate

Let

$$\bar{u}_\alpha = \int_Y u_\alpha(x) dx.$$

Our objective is now to derive an L^∞ estimate for

$$v_\alpha = u_\alpha - \bar{u}_\alpha.$$

Let λ be a constant to be chosen later, sufficiently large but fixed, and set

$$\phi_\alpha = e^{\lambda v_\alpha} + e^{-\lambda v_\alpha}.$$

Let r be a function with compact support in Q, independent of α, such that $0 \leqq r \leqq 1$, smooth and $r = 1$ on $\hat{Y}$.

We choose x to be the point of maximum of ϕ_α. Since ϕ_α is periodic, $x_\alpha \in Y$. Note that x_α is either a point of maximum of v_α or a point of minimum of v_α. We then fix G^α by (2.2). We shall prove the following main estimate

<u>Lemma 2.3</u> *One has*

$$\|u_\alpha - \bar{u}_\alpha\|_{L^\infty} \leqq C. \tag{2.6}$$

<u>Proof</u> We multiply (1.13) by the test function $(e^{\lambda v_\alpha} - e^{-\lambda v_\alpha})\, r^2 G^\alpha$ and integrate over Q. We deduce (using the summation convention)

$$\lambda \int a_{ij} \frac{\partial u_\alpha}{\partial x_j}\frac{\partial u_\alpha}{\partial x_i} \phi_\alpha\, r^2 G^\alpha\, dx$$

$$+ \int a_{ij} \frac{\partial u_\alpha}{\partial x_j} (e^{\lambda v_\alpha} - e^{-\lambda v_\alpha})(2r \frac{\partial r}{\partial x_i} G^\alpha + r^2 \frac{\partial G^\alpha}{\partial x_i})dx$$

$$+ \int a_i \frac{\partial u_\alpha}{\partial x_i} (e^{\lambda v_\alpha} - e^{-\lambda v_\alpha}) r^2 G^\alpha dx$$

$$= \int H\, (e^{\lambda v_\alpha} - e^{-\lambda v_\alpha})\, r^2 G^\alpha\, dx.$$

But

$$\int a_{ij} \frac{\partial u_\alpha}{\partial x_j} (e^{\lambda v_\alpha} - e^{-\lambda v_\alpha}) r^2 \frac{\partial G^\alpha}{\partial x_i} dx + \int a_j \frac{\partial u_\alpha}{\partial x_i} (e^{\lambda v_\alpha} - e^{-\lambda v_\alpha}) r^2 G^\alpha\, dx$$

$$= \frac{1}{\lambda} \int a_{ij} \frac{\partial}{\partial x_j} \phi_\alpha r^2 \frac{\partial G^\alpha}{\partial x_i} dz + \frac{1}{\lambda} \int a_i \frac{\partial \phi_\alpha}{\partial x_i} r^2 G^\alpha\, dx$$

$$= \frac{1}{\lambda} \int a_{ij} \frac{\partial}{\partial x_j} (\phi_\alpha r^2) \frac{\partial G^\alpha}{\partial x_i} dx + \frac{1}{\lambda} \int a_i \frac{\partial}{\partial x_i} (\phi_\alpha r^2) G^\alpha\, dx$$

$$- \frac{1}{\lambda} \int 2a_{ij}\, \phi_\alpha r \frac{\partial r}{\partial x_j} \frac{\partial G^\alpha}{\partial x_i} dx - \frac{1}{\lambda} \int 2a_i \phi_\alpha r \frac{\partial r}{\partial x_i} G^\alpha\, dx;$$

using this relation in (2.7), and taking account of the definition of the Green function (2.2), we obtain

$$\lambda \int a_{ij} \frac{\partial u_\alpha}{\partial x_j} \frac{\partial u_\alpha}{\partial x_i} \phi_\alpha r\, G^\alpha\, dx + \frac{1}{\lambda} \phi_\alpha(x_\alpha) + 2 \int a_{ij} \frac{\partial u_\alpha}{\partial x_i} (e^{\lambda v_\alpha} - e^{-\lambda v_\alpha}) r \frac{\partial r}{\partial x_j} G^\alpha$$

$$- \frac{2}{\lambda} \int a_{ij}\, \phi_\alpha r \frac{\partial r}{\partial x_j} \frac{\partial G^\alpha}{\partial x_i}\, dx - \frac{2}{\lambda} \int a_i \phi_\alpha r \frac{\partial r}{\partial x_i} G^\alpha\, dx \tag{2.8}$$

$$= \int H(e^{\lambda v_\alpha} - e^{-\lambda v_\alpha}) r^2 G^\alpha\, dx.$$

We notice that

$$(e^{\lambda v_\alpha} - e^{-\lambda v_\alpha}) \leqq \phi\alpha \quad .$$

Therefore we deduce from (2.8) that

$$\lambda\beta \int |Du_\alpha|^2 \phi_\alpha r^2 G^\alpha\, dx + \frac{1}{\lambda} \phi_\alpha(x_\alpha)$$

$$\leqq C \int_Q |Du_\alpha| \phi_\alpha r |Dr| G^\alpha\, dx + \frac{c}{\lambda} \int_Q \phi_\alpha r |Dr|\, |DG^\alpha|\, dx \tag{2.9}$$

$$+ \frac{c}{\lambda} \int_Q \phi_\alpha r |Dr| G^\alpha + K \int_Q \phi_\alpha\, G^\alpha r^2\, dx$$

$$+ K \int_Q |Du_\alpha|^2\, \phi_\alpha r^2 G_\alpha\, dx;$$

hence also

$$(\lambda\beta - K - \frac{c}{2}) \int_Q |Du_\alpha|^2\, \phi_\alpha r^2 G_\alpha\, dx + \frac{1}{\lambda} \phi_\alpha(x_\alpha)$$

$$\leqq \int_Q \phi_\alpha\, G^\alpha \left[\frac{c}{2} |Dr|^2 + \frac{c}{\lambda} r|Dr| + Kr^2\right] dx$$

$$+ \frac{c}{\lambda} \int_Q \phi_\alpha |Dg^\alpha| r |Dr|\, dx$$

$$\leqq C_1 \int_Q \phi_\alpha G^\alpha\, dx + C_2 \int_Q \phi_\alpha |DG^\alpha|\, dx$$

where C_1, C_2 are constants depending on λ, but λ is now fixed so that

$$\lambda\beta - K - \frac{c}{2} \geqq 1.$$

Let us write

$$\int_Q \phi_\alpha G^\alpha \, dx = \int_{Q\cap\{|v_\alpha|>h\}} \phi_\alpha G^\alpha \, dx + \int_{Q\cap\{|v_\alpha|<h\}} \phi_\alpha G^\alpha \, dx$$

$$\leqq \phi_\alpha(x\ ) \int_{Q\cap\{|v_\alpha|>h\}} G^\alpha \, dx + 2e^{\lambda h} \int_Q G^\alpha \, dx. \tag{2.11}$$

Similarly

$$\int_Q \phi_\alpha |DG^\alpha| \, dx \leqq \phi_\alpha(x_\alpha) \int_{Q\cap\{|v_\alpha|>h\}} |DG^\alpha| dx + 2e^{\lambda h} \int_Q |DG^\alpha| dx \tag{2.12}$$

and

$$\int_{Q\cap\{|v_\alpha|>h\}} G^\alpha \, dx \leqq \left[\int_Q (G^\alpha)^p dx\right]^{1/p} [\text{Meas}\{x \in Q | |v_\alpha|>h\}]^{1/q}$$

$$\int_{Q\cap\{|v_\alpha|>h\}} |DG^\alpha| dx \leqq \left[\int_Q |DG^\alpha|^p \, dx\right]^{1/p} [\text{Meas}\{x \in Q \big| |v_\alpha|>h\}]^{1/q} .$$

Choosing $1 < p < \frac{n}{n-1}$ the integrals of $(G^\alpha)^p$ and $|DG^\alpha|^p$ are bounded (see e.g. [3, 4]). Now

$$\text{Meas}\{|v_\alpha|>h\} \leqq \frac{1}{h^2} \int_Y v_\alpha^2 dx$$

$$= \frac{1}{h^2} \int_Y (u_\alpha - \bar{u}_\alpha)^2 \, dx \leqq \frac{c}{h^2} \int_Y |Du_\alpha|^2 \, dx$$

$$\leqq \frac{c}{h^2} .$$

Collecting results, we can assert from (2.11), (2.12)

$$\int_Q \phi_\alpha G^\alpha \, dx + \int_Q \phi_\alpha |DG^\alpha| \, dx \leqq C \frac{\phi_\alpha(x_\alpha)}{h^{2/q}} + Ce^{\lambda h}. \tag{2.13}$$

Using (2.13), (2.10) yields

$$\int_Q |Du_\alpha|^2 \phi_\alpha r^2 G^\alpha \, dx + \frac{1}{\lambda} \phi_\alpha(x_\alpha) \leqq C_1 \frac{\phi_\alpha(x_\alpha)}{h^{2/q}} + C_1 e^{\lambda h}.$$

But we can choose h sufficiently large, such that

$$\frac{1}{\lambda} > \frac{C_1}{h^{2/q}} .$$

Thus we have proved that

$$\phi_\alpha(x_\alpha) \leqq C.$$

Since x_α is a point of maximum of ϕ_α, it follows that

$$\phi_\alpha(x_\alpha) \leqq c, \forall x,$$

which implies

$$|v_\alpha(x)| \leqq C,$$

and yields the desired result.

2.3 Convergence

Let us set

$$z_\alpha(x) = u_\alpha(x) - u_\alpha(z_0).$$

From Lemma 2.2 and 2.3 we have

$$||z_\alpha||_{L^\infty} \leqq C, \quad |Dz_\alpha|_{L^2} \leqq C.$$

Pick a subsequence such that

$$z_\alpha \to z \text{ in } H^1 \text{ weakly and a.e.} \tag{2.14}$$

We shall prove that

<u>Lemma 2.4</u> $z_\alpha \to z$ *in* H^1 *strongly.*

<u>Proof</u> Consider $\chi_\alpha = e^{\lambda(z_\alpha - z)} - e^{-\lambda(\alpha_\alpha - z)}$ which is bounded (independently of α and periodic). Set also

$$\psi_\alpha = e^{\lambda(z_\alpha - z)} + e^{-\lambda(z_\alpha - z)},$$

with the equation (1.13) as

$$Az_\alpha + \alpha z_\alpha + \rho_\alpha = H(x, Dz_\alpha) \tag{2.15}$$

where

$$\rho_\alpha = \alpha u_\alpha(x_\alpha).$$

Without loss of generality we can assume that

$$\rho_\alpha \to \rho. \tag{2.16}$$

Note that from (2.14) we have

$$\chi_\alpha \to 0 \text{ a.e.} \tag{2.17}$$

We multiply (2.15) by χ_α and integrate over Y. We obtain

$$\lambda \int a_{ij} \frac{\partial}{\partial z_j} (z_\alpha - z) \frac{\partial}{\partial x_i} (z_\alpha - z)\psi_\alpha x + \lambda \int a_{ij} \frac{\partial z}{\partial x_j} \frac{\partial}{\partial x_i} (z_\alpha - z)\psi_\alpha dx$$

$$+ \int a_i \frac{\partial z_\alpha}{\partial x_i} \chi_\alpha \, dx + \int \alpha z_\alpha \chi_\alpha dx + \rho_\alpha \int \chi_\alpha \, dx \tag{2.18}$$

$$= \int H(x, Dz_\alpha)\chi_\alpha \, dx$$

$$\leqq K \int (1 + |Dz_\alpha|^2)|\chi_\alpha| \, dx$$

$$\leqq 2K \int |D(z_\alpha - z)|^2 \psi_\alpha \, dx + K \int (1 + 2|Dz|^2)|\chi_\alpha| \, dx.$$

We choose λ so that $\lambda\beta > 2K$. By virtue of (2.17), (2.14), and the L^∞ bound on z_α, we easily deduce from (2.18) that

$$\overline{\lim_{\alpha\to 0}} \int |D(z_\alpha - z)|^2 \, dx = 0$$

hence the desired result.

From the last assumption (1.4) we deduce

$$\int |H(x,Dz_\alpha) - H(x,Dz)|dx \leqq k_2 \int |D(z_\alpha - z)|(1+|Dz_\alpha|+|Dz|)dx \to 0 \text{ as } \alpha \to 0.$$

Therefore, we can pass to the limit in (2.15) and obtain

$$Az + \rho = H(x,Dz). \tag{2.19}$$

By a classical argument, the function z being H^1, bounded and solution of (2.19) is $W^{2,p}$. In fact, it is also true that z_α remains bounded in $W^{2,p}$, hence the convergence part is proved.

2.4 Uniqueness

Since the solution z is regular, its gradient is bounded, hence the quadratic growth does not matter any more. The uniqueness thus follows from standard maximum principle arguments. The proof of Theorem 1.1 has been completed.

Remark 2.1 The techniques apply in the case of the Dirichlet problem. For instance, the uniform elliptic equation

$$Au = -\sum_{i,j} \frac{\partial}{\partial x_i}\left(a_{ij}(x,u)\frac{\partial}{\partial x_j}\right) + a_0(x,u,Du) = 0$$

with $|a_0| \leqq k + K|Du|^2$ has a $C^\alpha \cap H_0^1$ solution if $(Au,u) \geqq c\,||u||^2_{H^1}$, $c > 0$. One has to approximate the problem via variational inequalities with the same operator A and the convex set $\mathbb{K} = \{u \in H_0^1, -L \leqq u \leqq L\}$, $L \to \infty$. This procedure does not destroy the coerciveness of the problems. Simple considerations concerning the admissibility of the test functions we used are involved.

REFERENCES

[1] Bensoussan, A.: Methodes de perturbation en controle optimal, to be published.

[2] Gimberg.: Thesis, Paris Dauphine 1984.

[3] Stampacchia, G.: Le problème de Dirichlet pour équations elliptiques du second ordre à coefficients discontinus. Ann. Inst. Fourier 15, 189-258 (1965).

[4] Widman, K.O.: The singularity of the Green function for nonuniformly elliptic partial differential equations with discontinuous coefficients. Technical report. Uppsala University 1970.

A. Bensoussan
Université de Paris-Dauphine
and INRIA
Paris
France

J. Frehse
Institut Angewandte Mathematik
Universität Bonn
5300 Bonn
West Germany

J P CECCONI

Problems of the calculus of variations in Banach spaces and classes BV

INTRODUCTION

I would like to thank the Organizing Committee for asking me to contribute to this Colloquium dedicated to the memory of the great mentor Leonida Tonelli. I, too, had the honour of being a student and disciple of Leonida Tonelli even though this unfortunately occurred during the War and Post-War period. The difficulties of that time led me to appreciate all the more his image of Master of life and science. I remember him with deep emotion.

Part A of this paper is concerned with certain problems of the calculus of variations, in view of existence theorems, in which the state variable has its values in a Banach space or Hilbert space. Part B is concerned, also in view of existence theorems, with problems in which the state variable has its values in $\mathbb{R}$, or $\mathbb{R}^n$, but is assumed to be a BV function in its domain. Corresponding references are listed in parts A and B of the bibliography. This choice of main topics is due to the fact that a team of researchers - that I have encouraged - in Genova, my present seat since the beginning of the sixties, has been particularly concerned with these two issues.

However, during the last twenty years many other problems concerning calculus of variations and control theory have been studied in Genova. To name but a few, contributions were made by T. Zolezzi both to problems of variational stability and to the theory of well posedness for problems of optimization (together with R. Lucchetti, F. Mignanego, F. Patrone). Results were also obtained by O. Caligaris, P. Oliva, and R. Peirone concerning problems of the calculus of variations of the Pareto type, and various other studies were carried out on the calculus of variations. These are not discussed here but some references are listed in parts C, D, E of the bibliography.

Thus, in part A, we are dealing with integrals $J(u,v) = \int_\Omega f(t,u(t),v(t))dt$, where $\Omega \subset \mathbb{R}^\nu$, $\nu \geq 1$, where u and v have their values in Banach spaces X, Y respectively, where v may or may not be related to u, and $f:\Omega\times X \times Y \to [-\infty, +\infty]$. Under assumptions of measurability and suitable conventions, then J also has its value in $[-\infty, +\infty]$. Note that we take here for X and Y general Banach spaces, while in most recent work X and Y are Sobolev spaces over Ω.

It is of interest to note that V. Volterra and J. Pérès in their 1936 book [VP] mentioned the use in functional analysis of abstract spaces with an abstract definition of limit (σ-limit spaces). J.L. Lions [L] has considered problems of evolution with the state variable taking its values in a general Banach space X, but his functionals of the calculus of variations are mostly norms and quadratic forms. In his recent book Cesari [C] (nos. 15.2 and 8.1) mentioned σ-limit spaces in connection with questions of semicontinuity.

A. PROBLEMS IN WHICH THE STATE VARIABLE HAS ITS VALUES IN A BANACH SPACE

1. As far as I know, studies on the calculus of variations in Banach spaces with integrands of general type started in the 70s. The simplest cases are those described by the following problems A and B.

Let X be a (real) Banach space, and let $J = [a,b] \subset \mathbb{R}$ be a closed finite interval on the reals. Given $p > 1$ let $W^{1,p}(J,X)$ denote the set of all functions $u : J \to X$ such that, for some function $v \in L^p(J,X)$ we also have $u(t) = u(a) + \int_a^t v(s)ds$ for all $t \in J$. This set can be made into a Banach space by means of any of the two equivalent norms

$$|u(a)| + \int_J |u'(t)|dt \quad \text{and} \quad |u|_{L^1(J)} + |u'|_{L^1(J)}$$

where of course $v = u'$ a.e. in J.

Let us consider an integrand $f : J \times X \times X \to [-\infty, +\infty]$ and denote by $\mathbb{A}$ a suitable subset of $W^{1,p}(J,X)$.

Problem A Let us discuss the existence of the minimum in $\mathbb{A}$ of the integral

$$I(u) = \int_a^b f(t,u(t),u'(t))dt, \quad u \in \mathbb{A},$$

where $u'(t) = v(t)$. Here the assumption is made that f is such that the map $t \in J \to f(t,u(t),u'(t)) \in [-\infty, +\infty]$ is measurable, and the integral is defined as the sum of the integrals of the positive and negative parts of $f(t,u(t),u'(t))$ with the convention that $(+\infty) + (-\infty) = +\infty$.

A wide range of problems of optimization can be reduced to the form above by deparametrization. This process gives rise to possible constraints on the values of the state u(t) and velocity u'(t), and these can be incorporated in the integrand, by allowing values $\pm\infty$ to f and to $I(u)$.

Only for the sake of simplicity we have worded this problem A in terms of simple integrals $I(u)$ (i.e., J is an interval in $\mathbb{R}$), instead of multiple integrals.

2. Now let X, Y be two (real) Banach spaces, and let Ω be a bounded open set in $\mathbb{R}$. Let us consider the Banach spaces $L_p(\Omega,X)$, $L_q(\Omega,Y)$ for fixed $p, q \geq 1$, a suitable subset $\mathbb{A}$ of $L_p(\Omega,X) \times L_q(\Omega,Y)$, and a given integrand $f : \Omega \times X \times Y \to [-\infty, +\infty]$.

Problem B Let us discuss the existence of the minimum in $\mathbb{A}$ of the integral

$$I(u,v) = \int_\Omega f(t,u(t),v(t))d\mu, \quad (u,v) \in \mathbb{A},$$

where μ is a finite measure in a bounded domain Ω of $\mathbb{R}^\nu$, $\nu \geq 1$. Here the assumption is made that f is such that the map $t\in\Omega \to f(t,u(t),v(t))\in[-\infty,+\infty]$ is summable for every $(u,v) \in \mathbb{A}$, and with the same conventions stated above. For the sake of simplicity we shall limit ourselves to the case $\nu = 1$ and $\Omega = J$ an interval in $\mathbb{R}$. For the sake of simplicity we assume here that μ is the Lebesgue measure in $\mathbb{R}$.

3. It should be noticed that problem A, unlike what happens in the finite dimensional case, cannot be reduced to problem (B). In fact, while for $\dim X < +\infty$

$$u_n \xrightarrow[W^{1,1}(J,R)]{} u_o \text{ is equivalent to } \left(u_n \xrightarrow[L^1(J,R)]{} u_o, \quad u'_n \xrightarrow[L^1(J,R)]{} u'_o\right),$$

instead, for $\dim X = +\infty$, only the following may occur:

$$u_n \xrightarrow[W^{1,1}(J,X)]{} u_o \text{ is equivalent to } \left(u_n(a) \xrightarrow[X]{} u_o(a), \ u'_n \xrightarrow[L^1(J,X)]{} u'_o\right),$$

as shown by the following example, which was presented by P. Oppezzi [21] for a different purpose. Indeed let X be a Hilbert separable space, $J = [0,1]$, and $(e_n)_{n\in N}$ be a fundamental system for X. For each $n \in N$ let

$$u_n : t \in [0,1] \to u_n(t) = te_n \in X,$$

and also

$$u_o : t \in [0,1] \to u_o(t) = 0 \in X.$$

Then, obviously we obtain

$$u_n \xrightarrow[W^{1,1}(J,X)]{} u_o$$

and $\qquad u_n \not\rightharpoonup u_o$ weakly in $L_1(J,X)$.

4. Let us consider now *Problem A* under the assumptions that X is reflexive and separable, and p = 1. Let us assume that the integrand f be normal in the sense that

(i) f is measurable with respect to $\mathcal{L}(J) \otimes B(X,X)$ generated by the Cartesian products of Lebesgue measurable sets of J times Borel measurable sets of $X \times X$, and $J = [0,1]$

(ii) $(s,v) \in X \times X \to f(t,s,v) \in [-\infty, +\infty]$ is weakly sequentially lower semicontinuous for almost all $t \in I$.

(iii) $v \in X \to f(t,s,v) \in [-\infty, +\infty]$ is convex for almost every $t \in I$ and for all $s \in X$.

Then it is known that $I(x)$ is defined for every $x \in W^{1,2}(J,X)$.

Now let us assume that f satisfies the Tonelli-Nagumo growth condition:

(iv) there is $\Phi : [0, +\infty[\to [0, +\infty[$ such that

$$\Phi(y)/y \to +\infty \text{ as } y \to +\infty, \text{ and } f(t,s,v) \geq \Phi(|v|)$$

for all $(t,s) \in J \times X$;

and also the following further assumptions:

(v) there is some $\bar{u} \in \mathbb{A}$ such that $I(\bar{u}) \in \mathbb{R}$;

(vi) $\mathbb{A} = W^{1,1}(J,X) \cap \{u(a) \in K\}$, where K is a weakly compact part of X.

Under hypotheses (i) - (vi), the integral $I(u)$ of problem A attains an absolute minimum at some $u_o \in \mathbb{A}$ with $I(u_o) \in \mathbb{R}$.

Condition (iv) above can be replaced by the following one

(iv') for every $r, \rho \in R_+$ there is some function $\phi_{r,\rho} \in L^1(J)$ such that $\langle p,v\rangle - f(t,s,v) \leq \phi_{r,\rho}(t)$ for all $|s| \leq r$, $|p| \leq \rho$, $v \in X$, and almost all $t \in J$. (For conditions of this type see Cesari [C], no. 10.4).

5. At first sight the weak sequential lower semicontinuity assumption (ii) for $f(t,.,.)$ may seem needlessly restrictive. However, this hypothesis cannot be easily removed and still guarantees the existence of the minimum as the following example by Caligaris and Oliva [15] shows. Let X be a separable Hilbert space, let $(e_n)_{n \in N}$ be an orthonormal base for X, and take

$$f(t,s,v) = |v|^2 + (|s|^2-1)^2 + \sum_{n=1}^{\infty} 2^{-n}\langle s,e_n\rangle^2,$$

$$\mathbb{A} = W^{1,1}(J,X) \cap (|u(0)| = 1), \ J = [0,1].$$

For the sequence $u_n(t) = e_n$ for $n \in N$, and $u_0(t) = 0$, $t \in I$, we have now

$$u_n \xrightarrow[W^{1,1}(J,X)]{} u_0,$$

$$f(t,u_n(t), u_n'(t)) = 2^{-n}, \ n = 1,2,\ldots$$

$$f(t,u_0(t), u_0'(t)) = 1, \quad I(u_n) = 2^{-n}, \ I(u_0) = 1.$$

Hence,

$$1 = I(u_0) > \varliminf_{n\to\infty} I(u_n) = 0,$$

and $I(u)$ is not weakly sequentially lower semicontinuous in $\mathbb{A}$. Furthermore,

$$\inf (I(u), u \in \mathbb{A}) = 0,$$

but there is no $\bar{u} \in \mathbb{A}$ for which $I(\bar{u}) = 0$. Indeed, $I(\bar{u}) = 0$ would imply $\int_0^1 |\bar{u}'(s)|^2 ds = 0$, hence, $\bar{u}(s) = c$ a.e. in J for some constant c, and $\int_0^1 \sum_{n=1}^{\infty} 2^{-n}\langle c,e_n\rangle^2 ds = 0$, hence $\langle c,e_n\rangle = 0$ for all $n \in N$, and $c = 0$. Finally, $\int_0^1 (|\bar{u}(t)|^2-1)^2 dt = 1$, and this contradicts $I(\bar{u}) = 0$. It is evident that $f(t,s,v)$ is strongly continuous in the variable (t,s,v), and possesses Fréchet derivatives of all orders. Hence, f is of the Carathéodory type in $J \times X \times X$, and thus $\mathcal{L}(J) \times B(X \times X)$ is measurable. Furthermore the function $v \in X \to f(t,s,v) \in [-\infty, +\infty]$ is convex, and from $s_n \xrightarrow[X]{} s_0$, $v_n \xrightarrow[X]{} v_0$, it follows that

$$f(t,s_0,v_0) \leq \underline{\lim}_{n\to\infty} f(t,s_n,v_n).$$

Finally, f satisfies the Tonelli-Nagumo growth condition. However, the weakly sequential lower semicontinuity hypothesis is not satisfied. In fact,

$$(e_n,0) \xrightarrow[X\times X]{} (0,0), \quad f(t,e_n,0) = 2^{-n}, \quad n = 1,2,\ldots,$$

$$\underline{\lim}_{n\to\infty} f(t,e_n,0) = 0 < f(t,0,0) = 1.$$

6. Caligaris and Oliva ([15], [16], [17]) obtained existence theorems for the minimum of abstract control problems of the following general type. Let X, Y be reflexive separable Banach spaces, let K be an integrand function, $K : J \times X \times Y \to [-\infty, +\infty]$, $J = [0,1]$, and let F be a given functional $F : W^{1,1}(J,X) \to \mathbb{R}$. We consider the problem of the minimum of the new functional

$$I(x,u) = F(x) + \int_0^1 K(t,x(t),x'(t),u(t))dt, \; (x,u) \in W^{1,1}(J,X) \times B(J,Y),$$

where B(J,Y) is the space of all Borel functions on [0,1] with values in Y, and with the usual conventions concerning measurability and the value of the integral. With suitable assumptions on K, this problem can be reduced to one of the free type considered above by means of the process of deparametrization. Indeed, by introducing the function

$$L(t,x,v) = \text{Inf}\ \{K(t,x,v,u),\ u \in Y\},$$

we have the new problem of the minimum of the functional

$$H(x) = F(x) + \int_0^1 L(t,x(t),x'(t))dt, \; x \in W^{1,1}(J,X).$$

For instance, for the problem of optimal control concerning the minimum of

$$\int_0^1 \psi(t,x(t),u(t))dt, \; (x,u) \in W^{1,1}(J,X) \times B(J,Y),$$

with constraints

$$x(t) \in X(t), \; u(t) \in U(t,x(t)), \; x'(t) = f(t,x(t),u(t)), \; t \in J(\text{a.e.}),$$

we define K(t,x,v,u) by taking

$$K(t,x,v,u) = \begin{cases} \psi(t,x,u) \text{ if } x \in X(t),\ u \in U(t,x),\ v = f(t,x,u), \\ +\infty \text{ otherwise.} \end{cases}$$

7. J.L. Lions [L] has studied a number of variational problems in connection with a Banach space X and a Hilbert space H with $X \hookrightarrow H \hookrightarrow X^*$, where X^* is the dual of X and both inclusions are everywhere dense and continuous. Ada Bottaro Aruffo [1], [2], [3] has studied a problem of such type monitored by a linear parabolic equation

$$x'(t) = A(t)x(t) + B(t)u(t) + f(t). \qquad (*)$$

For fixed real S,T, $S < T$, let us consider the space $W^p(S,T) = \{x \in L^p(S,T);X)$, $x' \in L^{p'}(S,T;X^*)\}$ where $p \geq 2$, $1/p + 1/p' = 1$, and take $f \in L^r(S,T;H)$ with $p' \leq r \leq q$. For every $t \in [S,T]$ let $A(t) : X \to X^*$ and $B(t) : U \to H$, be given operators, and where U is a given reflexive and separable Banach space. Let $F : [S,T] \times H \times U \to [-\infty, +\infty]$ be a given integrand, which we assume to be $\mathcal{L}(S,T) \times B(X \times U)$ measurable, and such that $F(t,\cdot,\cdot)$ is weakly sequentially lower semicontinuous. Let $g : H \times H \to [-\infty, +\infty]$ be a given weakly sequentially lower semicontinuous map, and let us study the problem of the minimum of the functional

$$G(x,u) = \int_S^T F(t,x(t),u(t))dt + g(x(S), x(T))$$

in a set C of pairs $(x,u) \in W^p(S,T) \times L^q(S,T;U)$ satisfying the differential equation (*) a.e. in [S,T]. We shall assume that C is weakly sequentially closed. Furthermore, we assume that A and B satisfy measurability conditions on [S,T], and that for every $t \in [S,T]$ either $A(t) \in L_0(X,X)$ or A(t) is a coercive and hemicontinuous bounded operator, besides further assumptions on B and growth conditions on F and g. Under the above hypotheses, Ada Bottaro Aruffo has proved existence theorems for the minimum of G, which extend previous results of V. Barbu in case of convex functionals, and through an ad hoc method of deparametrization. Her results include also existence and uniqueness theorems for solutions of the differential equation (*) under various hypotheses on A and B.

8. We wish to end our survey on problem A by mentioning a result of Oppezzi

[21] on the required convexity of the function

$$v \in X \to f(t,s,v) \in [-\infty, +\infty],$$

in connection with the sequential semicontinuity of the functional $I(x)$ and the extension of theorems of Tonelli, Cesari, et al. (cf. [C]) to the case where the state variable has its values in a Banach space. More specifically, let X be a Banach space, and let us assume that the integrand $f(t,s,v)$, or $f : J \times X \times X] \to [-\infty, +\infty]$ is $\mathcal{L}(J) \otimes B(X \times X)$ measurable, such that the function $s \in X \to f(t,s,v) \in [-\infty, +\infty]$ is continuous for almost every t and for every $v \in X$, and the function $v \in X \to f(t,s,v) \in [-\infty, +\infty]$ is lower semicontinuous for almost every $t \in J$ and for every $s \in X$. Under these hypotheses, Oppezzi has proved that, if the functional

$$u \in W^{1,\infty}(J,X) \to I(u) = \int_J f(t,u(t),u'(t))dt \in [-\infty, +\infty],$$

defined with the usual conventions, is sequentially lower semicontinuous with respect to a suitable weak topology in $W^{1,\infty}(J,X)$ (different from $\sigma(W^{1,\infty}(J,X), W^{1,1}(J,X^*))$, then under an appropriate growth hypothesis on f, the function

$$v \in X \to f(t,s,v) \in [-\infty, +\infty]$$

is necessarily convex for almost every $t \in J$ and for every $s \in X$.

9. Let us now move on to problem B and let us study the functional

$$I(x,u) = \int_\Omega f(t,x(t),u(t))dt$$

defined for $(x,u) \in L^p(\Omega,X) \times L^q(\Omega,Y)$ where X,Y are given Banach spaces. Bottaro and Oppezzi [11], [12], [13] have proved necessary and sufficient conditions for $I(x,u)$ to be semicontinuous with respect to the strong convergence of x in $L^p(\Omega,X)$ and to the weak convergence of y in $L^q(\Omega,Y)$ with $p \geq 1$, $q > 1$. These results extend previous ones by I.D. Ioffe for the case in which x and u have their values in Cartesian spaces.

For instance, these are the assumptions in one of the situations considered by Bottaro and Oppezzi.

(A) Let X,Y be separable reflexive Banach spaces, let Ω be a locally compact topological space which is σ-compact, let $B(\Omega)$ denote the Borelian family in Ω, let $\mathcal{L}$ be a σ-algebra of subsets of Ω containing $B(\Omega)$, and let $\mu:\mathcal{L} \to [0,+\infty]$

be a regular and complete measure.

(B) Let us assume that the integrand $f : \Omega \times X \times Y \to [-\infty , +\infty]$ satisfies a slightly more general measurability hypothesis than $\mathcal{L} \otimes B(X \times Y)$ measurability, namely, f is an almost Borel function in the sense that there is some $\Omega' \subset \Omega$, $\Omega' \in B(\Omega)$, with $\mu(\Omega \times \Omega') = 0$ and a $B(\Omega \times X \times Y)$ measurable function $\bar{f} : \Omega \times X \times Y \to [-\infty, +\infty]$, such that

$$f|_{\Omega' \times X \times Y} = \bar{f}|_{\Omega' \times X \times Y}.$$

(C) (i) $f(t,\cdot,\cdot)$ is sequentially lower semicontinuous with respect to the product topology of the strong topology of X times the weak topology of Y for almost every $t \in \Omega$.

(ii) $f(t,s,\cdot)$ is convex for almost every $t \in \Omega$ and for every $s \in X$.

It is further assumed that for at least one pair $(\bar{x},\bar{u}) \in L^p(\Omega,X) \times L^q(\Omega,Y)$ we have $J(\bar{x},\bar{u}) \in \mathbb{R}$.

Under these assumptions, Bottaro and Oppezzi proved that the functional $I(x,u)$ is sequentially lower semicontinuous with respect to the product topology obtained from the strong topology of $L^p(\Omega,X)$ times the weak topology of $L^q(\Omega,Y)$ if and only if the following equivalent conditions (a) and (b) are satisfied.

Condition (a): f satisfies the lower compactness property in $L^p(\Omega,X) \times L^q(\Omega,Y)$, that is, for every sequence $(x_n,u_n) \in L^p(\Omega,X) \times L^q(\Omega,Y)$ with

$$x_n \xrightarrow[L^p(\Omega,X)]{} x_0, \qquad u_n \xrightharpoonup[L^q(\Omega,Y)]{} u_0,$$

and for which the sequence $(I(x_n,u_n))_{n\in N}$ is bounded above, the following sequence of functions

$$t \in \Omega \to f[t,x_n(t),u_n(t)] \wedge 0 \in [-\infty, 0]$$

is weakly relatively compact in $L^1(\Omega,\mathbb{R})$.

Condition (b): there are functions $\bar{p} : \Omega \times X \to Y$ almost Borelian and $b \in L^1(\Omega,\mathbb{R})$, and a positive constant c such that

(b.1) $f(t,s,v) \geq \langle v,\bar{p}(t,x)\rangle - c|x|^p + b(t)$ for every $(t,s,v) \in \Omega \times X \times X$;

(b.2) for every convergent sequence $(x_n)_{n\in N}$ in $L^p(\Omega,X)$ which is weakly convergent in $L^q(\Omega,Y)$ and there exists $a \in \mathbb{R}$ such that $J(x_n,u_n) \leq a$ for all $n \in N$,

$$|\bar{p}(\cdot,u_n)|^{q'}, \quad n \in N,$$

is weakly relatively compact in $L^1(\Omega,\mathbb{R})$.

Actually Condition (b) is a refined version of the rougher condition (a) which is still valid as a necessary and sufficient condition for lower semicontinuity when $L^1(\Omega,X)$ and $L^q(\Omega,Y)$ with strong and weak topologies respectively, are replaced by two decomposable spaces L, M, i.e. L, M are such that, if $E \subset \Omega$, $v_1,v_2 \in L$ (or $v_1,v_2 \in M$), then $\chi_E v_1 + (1 - \chi_E)v_2 \in L$ (or M), where χ_E is the characteristic function of E in Ω, and $L \subset L^p(\Omega,X)$, $M \subset L^q(\Omega,Y)$ have properties similar to those of $L^p(\Omega,X)$, $L^q(\Omega,Y)$. (For details and proofs we refer to Bottaro and Oppezzi [13]).

10. Bottaro and Oppezzi [12] have also proved that, if hypothesis (A) in the above mentioned theorem remains unaltered, then conditions (B), (C), besides being sufficient for the sequential lower semicontinuity of J(x,u) with respect to the topology resulting from the product of the strong topology of $L^p(\Omega,X)$ times the weak topology of $L^q(\Omega,Y)$, are also necessary. In other words, they have proved that, if $f: \Omega \times X \times Y \to [-\infty, +\infty]$ is such that for every $x \in L^1(\Omega,X)$, $u \in L^1(\Omega,Y)$ the function $t \in \Omega \to f(t,x(t),u(t)) \in [-\infty,+\infty]$ is measurable, and if

$$J(x,u) = \int_\Omega f(t,x(t),u(t))d\mu(t)$$

is sequentially lower semicontinuous in $L^p(\Omega,X) \times L^q(\Omega,Y)$, then there is a function $\tilde{f} : \Omega \times X \times Y \to [-\infty, +\infty]$, which is $L(\Omega) \otimes B(X \times X)$-measurable, and such that

(1) $\tilde{f}(t,x(t),u(t)) = f(t,x(t),u(t))$ for almost all $t \in \Omega$, and for every pair (x,u) of measurable functions $x : \Omega \to X$, $u : \Omega \to Y$;

(2) $\tilde{f}$ verifies the lower semicontinuity and convexity properties stated in (C).

Bottaro and Oppezzi have also proved that the same conclusions are still valid if $L^p(\Omega,X)$, $L^q(\Omega,Y)$ are replaced by appropriate decomposable spaces L, M such that

$$L^{\infty}(\Omega,X) \subset L \subset L^{1}(\Omega,X), \quad L^{\infty}(\Omega,Y) \subset M \subset L^{1}(\Omega,Y).$$

For details and proofs we refer to Bottaro and Oppezzi [12].

Subsequently, C. Castaing and L. Clauzure [27] proved sufficient conditions for the lower semicontinuity of functionals of the type J(x,u) by considering an even more general setting consisting of a couple of Suslinian spaces X, Y, with the latter space Y σ-compact and locally convex, and such that its dual Y^{*} admits of a locally convex Suslinear topology compatible with the duality. It should be noticed that their proofs of the lower semicontinuity of J(x,u) use similar techniques (particularly the Yosida regularization), and their results, for the situations taken into consideration, are particular cases of those obtained by Bottaro and Oppezzi. The same can be said for some very recent results obtained by J. Balder [23] in a setting similar to that considered by Castaing and Clauzure [27]. An interesting detail in Balder's work is the fact that no a priori measurability hypotheses are assumed on the integrand since the functional values are exterior integrals.

11. The problem of an integral representation of local functionals F(u,A) depending on a function f and a set A was discussed by Bottaro and Oppezzi in connection with Banach spaces.

If, for example, F(u,A) is defined in $W^{1,1}(\Omega,\mathbb{R}) \times B(\Omega)$ the problem of an integral representation consists in finding conditions on F(u,A) under which there exists some integrand f(t,s,v) such that

$$F(u,A) = \int_A f(t,u(t), (\operatorname{grad} u)(t))dt$$

for every $(u,A) \in W^{1,1}(\Omega,\mathbb{R}) \times B(\Omega)$. A complete solution to this problem was recently provided by G. Buttazzo and G. Dal Maso [26]. Recent work by Bottaro and Oppezzi [14], [22] concerns the general problem of integral representation for functionals depending on functions belonging to Sobolev spaces $W^{m,p}(\Omega,X)$ for $1 \leq p \leq +\infty$ and where X is a Banach space. For the sake of brevity, we consider here only the case $1 \leq p < +\infty$. Given a separable Banach space X, a bounded open set Ω of $\mathbb{R}^{\nu}$, a positive integer m, and a real number $p > 1$, let us denote by $W^{m,p}(\Omega,X)$ the Banach space of all functions $u \in L^{p}(\Omega,X)$ with derivatives (in the weak sense) of every order α, $|\alpha| \leq m$, with $D^{\alpha}u \in L^{p}(\Omega,X)$, and norm

$$|u|_{m,p,\Omega} = \Sigma_{|\alpha|\leq m} |D^{\alpha}u|_{p,\Omega}.$$

Now let us consider the classes $\mathbb{A}(\Omega)$, $\mathbb{A}_0(\Omega)$ of all open sets in Ω (respectively, of all open sets in Ω with compact closure) and the functional

$$F : (u,A) \in W^{m,p}(\Omega,X) \times B(\Omega) \to F(u,A) \in [-\infty, +\infty],$$

and let us introduce the following definitions.

(1) We say that F is a measure if for every $u \in W^{m,p}(\Omega,X)$ the map

$$B \in B(\Omega) \to F(u,B) \in [-\infty, +\infty]$$

is a measure.

(2) We say that F is strongly lower semicontinuous if for every $A \in \mathbb{A}_0(\Omega)$, and u_n, $u \in W^{m,p}(\Omega,X)$ with $|u_n-u|_{m,p,A} \to 0$, we also have

$$F(u,A) \leq \varliminf_{n\to\infty} F(u_n,A).$$

(3) We say that F is p-bounded if there is some $\phi \in L^1_{loc}(\Omega,\mathbb{R})$ such that for every $(u,A) \in W^{m,p}(\Omega,X) \times B(\Omega)$ we have

$$F(u,A) \leq \int_A [\phi + c\, \Sigma_{|\alpha|\leq m} |D^{\alpha}u|^p]dt.$$

(4) Having set $Z = \prod_{r=0}^{m} X^{\binom{n+2-1}{r}}$, we say that a Carathéodory function $f : \Omega \times Z \to \mathbb{R}$ has a growth condition of order p if there exist $\phi \in L^1_{loc}(\Omega,\mathbb{R})$ and $C \in \mathbb{R}_+$ such that $|f(t,z)| \leq \phi(t) + C|z|^p$.

(5) Let X_0 be a countable part of X dense in X; let $P_0 = \{p : \mathbb{R}^n \to \mathbb{R}$, p a polynomial with coefficients in X_0; and take

$Q(t,r) = \prod_{i=1}^{n} (t_i - \frac{r}{2}, t_i + \frac{r}{2})$ for every $t = (t_1,t_2,\dots,t_n) \in \mathbb{R}$ and $r \in \mathbb{R}_+$. We say that F has the $\alpha(P_0)$ property if for every $u \in W^{m,p}_{loc}(\Omega,X)$ there exists $\Omega(u_0) \subset \Omega$, $|\Omega\setminus\Omega_0| = 0$, such that for every $t \in \Omega(u_0)$, M, $\delta \in \mathbb{R}_+$ there are $\sigma(u_0,\delta,M,t)$, $a(u_0,\delta,M,t) \in \mathbb{R}_+$ such that for any $0 < a < a(u_0,\delta,M,t)$ we have

$$a^{-n}|F(u_0 + u, Q(t,a)) - F(u_0 + v, Q(t,a)| < \delta$$

when

$$|u-v|_{m,\infty,Q(t,a)} \leq \sigma(u_0,\delta,M,t),$$

and

$$|u|_{m,\infty,Q(t,a')},\ |v|_{m,\infty,Q(t,a')} \leq M \quad \text{for at least one } a' > a.$$

With these definitions, we can now state the following representation theorem: If

$$F : (u,A) \in W^{m,p}_{loc}(\Omega,X) \times B(\Omega) \to [-\infty, +\infty]$$

is a measure, is strongly lower semicontinuous, is p-bounded, and has the $\alpha(P_0)$ property, then there exists a Carathéodory integrand $f : \Omega \times Z \to [-\infty,+\infty]$ satisfying a growth condition of order p such that

$$F(u,A) = \int_A f(t,(Du)(t))dt,$$

for every $(u,A) \in W^{m,p}_{loc}(\Omega,X) \times \mathbb{A}_0(\Omega)$. (Bottaro, Oppezzi).

A similar result holds for $p = \infty$. Also, the question of the existence of the $\Gamma(\tau^-)$-limit $F_n(\cdot,A)$ has been investigated, together with the representation theorems for this Γ-limit. Hence, a representation theorem can be obtained for the functional $sc^-\cdot F(\cdot,A)$, which is the lower semicontinuous envelope of the integral functional $F(\cdot,A)$.

12. To end this first part A, we wish to report on extensions of classical theorems concerning composition operators, or Nemitskii -operators. In other words, given a function $f : \Omega \times \mathbb{R} \to \mathbb{R}$ of the Carathéodory type, we consider the map

$$N_f : u \in L^p(\Omega,\mathbb{R}) \to N_f(u) = f(\cdot,u),$$

and we ask whether N_f is of class $L^q(\Omega,R)$. Results of this type have been obtained by R. Lucchetti and F. Patrone [20] for f $\mathcal{L}(\Omega) \otimes B(X)$-measurable integrand, and X and Y separable Banach spaces. Further extensions have been obtained by A. Bottaro Aruffo [8], [9] for the following situation.

Let $g_1,g_2 : \Omega \times X \to [-\infty, +\infty]$ be given $\mathcal{L}(\Omega) \otimes B(X)$-measurable maps, and let $f : \Omega \times X \to Z$ be a $\mathcal{L}(\Omega) \times B(X)$-measurable map. The question is whether, for every measurable function $u : \Omega \to X$ with $g_1(\cdot,u) \in L^1(\Omega,\mathbb{R})$, $g_2(\cdot,u) \in$

$L^\infty(\Omega,R)$, we also have that $f(\cdot,u) \in L^q(\Omega,X)$.

If $g_1(\cdot,x) = |x|_X^p$, $g_2(\cdot,x) = 0$, and $X = \mathbb{R}$, then we are in the classical case.

In the more general context mentioned above, A. Bottaro Aruffo [8] has proved that, if for example, the domain of N_f is $L^p(\Omega,X)$, then f satisfies suitable inequalities which, in the case where $g_1(\cdot,x) = |x|_X^p$, $g_2(\cdot,x) = 0$, fall within the classical ones. Lucchetti and Patrone have already communicated this result.

As for the continuity property of N_f, A. Bottaro Aruffo [9] proved that, within the context that she considered, f is necessarily a Carathéodory function. This result, when stated for example in the simple context where

$$g_1(\cdot,x) = |x|_X^p, \quad g_2(\cdot,x) = 0, \quad X \text{ a Banach space},$$

asserts that, if N_f is continuous from the strong topology of $L^p(\Omega,X)$ to the strong topology of $L^q(\Omega,X)$, then f is of the Carathéodory type, in the sense that, besides the measurability of $f(\cdot,x)$ for every $x \in X$, we also have

$$f(t,\cdot) \in C((X,\|\cdot\|_X), (Z,\|\cdot\|_Z))$$

for almost every $t \in \Omega$. Analogously, if N_f is continuous from the strong topology of $L^p(\Omega,X)$ to the weak topology of $L^q(\Omega,X)$, then f is of the Carathéodory type in the sense that, besides the measurability of $f(\cdot,x)$ for every $x \in X$, we also have

$$f(t,\cdot) \in C((X,\|\cdot\|_X), (Z, \text{weak topology of } Z))$$

for almost every $t \in \Omega$. This result extends a previous one of G. Buttazzo and G. Dal Maso [26] in the case of finite dimensional spaces.

13. A Bottaro Aruffo has applied the results above to the study of relaxation for the problem of the minimum of the integral

$$\int_\Omega f[t,G(p(t)), p(t)]d\mu(t), \quad p \in \mathbb{A} \subset L^1(\Omega,Y),$$

that is, to the study of the minimum of the functional

$$\text{sc} - \int_\Omega f[t,G(p(t)), p(t)]d\mu(t), \quad p \in \mathbb{A},$$

for a topology on $\mathbb{A}$ which will be described below, and where f is a normal integrand $f : \Omega \times X \times Y \to [-\infty, +\infty]$, where X is a metrizable Suslinian topological space, and Y is a reflexive separable Banach space with the weak topology.

In this context, and where moreover growth conditions of f are expressed as integral dominations, A. Bottaro Aruffo [18], using previous results concerning equivalence between integral domination and uniform integrability [7], has obtained a result concerning relaxation. Namely, in the case where X, Y are reflexive separable Banach spaces, the result reduces to the following one, which was suggested by previous theorems by R. Temam and I. Ekeland concerning finite dimensional spaces.

Let ϕ be a lower semicontinuous convex increasing function ϕ: $[0, +\infty[\to [0, +\infty[$ such that $\phi(s)/s \to +\infty$ as $s \to +\infty$. Let Ω be a locally compact metric space, let $\mathcal{L} \supset B(\Omega)$ be a σ-algebra, and let μ be a complete separable measure with no atoms. Let us consider the subset of $L^1(\Omega,Y)$ defined by

$$L^\phi(\Omega,Y) = \{p \in L^1(\Omega,Y) : \int_\Omega \phi(|p(t)|)d\mu(t) < +\infty\} ,$$

let us consider the normal integrand $f : \Omega \times X \times X \to [-\infty, +\infty]$ for the product topology of the strong topology on X times the weak topology of Y, and let us consider the further normal integrand $g : \Omega \times Y \to [0, +\infty]$ for the weak topology of Y. Let us assume that there is $a \in L^1(\Omega,R)$ such that

$$g(t,v) + a(t) \leqq f(t,s,v) \text{ for almost every } t \in \Omega, \text{ and all } (s,v) \in X\times Y,$$

$$\phi(|v|_y) \leqq g(t,v) \text{ for almost every } t \in \Omega \text{ and all } v \in Y.$$

Given $\beta > 1$ and a transformation $G : L^\phi(\Omega,Y) \to L^\beta(\Omega,X)$ which is (ϕ,β)-compactifying, in the sense that, if $p_n \in L^1(\Omega,X)$ for every $n \in N$ with $p_n \to p$ in $\sigma(L^1(\Omega,X), L\ (\Omega,X^*))$, and

$$\sup_{n\in N} \{\int_\Omega \phi[|p_n(t)|]d\mu(t) < +\infty,$$

then

$$|G(p_n) - G(p)|_{L^p(\Omega,X)} \to 0.$$

Given $1 \leqq \beta < \infty$, let us suppose that there are $C_o \in \mathbb{R}_+$ and $b_o \in L^1(\mu,\mathbb{R})$ such that

$$f(t,s,v) \leqq b_o(t) + C_o(|s|_X^B + G(t,v))$$

for almost every $t \in \Omega$ and all $(s,v) \times X \times Y$. Finally, let us suppose that there are $C \in \mathbb{R}_+$, $b \in L^1(\mu,\mathbb{R})$, $\Psi:[0, +\infty) \to \mathbb{R}_+$, bounded in the bounded parts of $[0, +\infty)$, such that $\phi(t)/\Psi(t) \to \infty$ as $t \to +\infty$, and such that we have

$$|f(t,s,v) - f(t,s',v')| \leqq b(t) + C|s|^\beta + |s'|^\beta + \Psi(|v|),$$

for almost every $t \in \Omega$, all $s, s' \in X \times X$, and all $v,v' \in Y$ for which $f(t,s,v), f(t,s',v') \in \mathbb{R}$.

Under the assumptions above (and also under more general ones), A. Bottaro Aruffo proved that

$$sc^- \int_\Omega f(t,G(p(t)), p(t))d\mu(t)$$

$$= \int_\Omega (f(t,G(\alpha(t)),\cdot)^{**} p(t)d\mu$$

where sc^- is taken with respect to the topology $\sigma(L^1(\Omega,Y), L^\infty(\Omega,Y^*))$. Moreover, she proved that the relaxed problem attains a minimum in $L^\phi(\Omega,Y)$, and that every solution of this problem is a limit of minimizing sequences of the given problem when $\mathbb{A} = L^\phi_-(\Omega,Y)$. Similar results hold for $p = \infty$.

B. PROBLEMS IN WHICH THE STATE VARIABLE IS A REAL VALUED FUNCTION OF CLASS BV

14. Let us now deal with functionals in the calculus of variations for which the state is a BV function in its domain.

But first let us go back to problem (A) considered above. It is well known that in the hypothesis of continuity of the integrand $f(t,s,v)$ and of convexity of f in the variable v, there is generally no minimum for

$$I(u) = \int_\Omega f[t,u(t),u'(t)]dt, \quad u \in W^{1,1}(\Omega), \quad \Omega = [0,1],$$

if f does not have a growth more than linear in $|v|$. Assuming, for example

$$g(t,s) = \begin{cases} |1-|s||(\tfrac{1}{2}-t) & \text{if } 0 \leqq t \leqq 1/2, \\ |s|(t-\tfrac{1}{2}) & \text{if } \tfrac{1}{2} \leqq t \leqq 1 \end{cases} \qquad (t,s) \in [0,1] \times \mathbb{R},$$

$$f(t,s,v) = g(t,s) + |v|/9.$$

It can be easily seen that

$$\inf \{I(u) = \int_\Omega f(t,u(t), u'(t))dt, \ u \in W^{1,1}(\Omega)\} = 1/9,$$

while, for $u \in W^{1,1}(\Omega)$, we have $I(u) > 1/9$.

On the other hand, it is also known that, if Ω is an open bounded set in $\mathbb{R}^n$ with boundary of class $C^{1,0}$, then, in general, there is no solution for the problem concerning the minimum of the integral

$$I(u) = \int_\Omega \{1 + |\text{grad } u|^2\}^{1/2}dt, \ u \in W^{1,1}(\Omega), \ u|_{\partial\Omega} = \phi \in L^1(\partial\Omega).$$

(nonparametric problem of the area), while it is well known that if the same problem is set in $BV(\Omega)$, rather than in $W^{1,1}(\Omega)$, it has solutions if the functional $I(u)$ is properly extended to $BV(\Omega)$.

These considerations (for the moment we continue referring to the 1-dimensional case with $\Omega = [0,1] \subset \mathbb{R}$) lead to the extension of the functional $I(u)$ to $BV(\Omega)$ in the lower semicontinuous way, as it has already been done, for example, by J. Serrin, by setting

$$J(u) = \inf \{\underline{\lim}_{n\to\infty} J(u_n), \ u_n \in W^{1,1}(\Omega), \ u_n \xrightarrow[L^1(\Omega)]{} u\},$$

or by replacing $L^1(\Omega)$ with $L^1_{loc}(\Omega)$ for every $u \in BV(\Omega)$.

This procedure does not allow for an exhaustive discussion of the behaviour of u at the extremes of the interval $\Omega = [0,1]$; this is fundamental for many problems of the calculus of variations.

15. It is perhaps more interesting to proceed with a different extension. While continuing to refer to the 1-dimensional case, it should be kept in mind that $W^{1,1}(\Omega)$ can be identified with $\mathbb{R} \oplus L^1(\Omega)$ through

$$u \in W^{1,1}(\Omega) \leftrightarrow \{u(0), u'\} \in \mathbb{R} \oplus L^1(\Omega),$$

so that the norms

$$|u|_{L^1(\Omega)} + |u'|_{L^1(\Omega)}, \quad |u(0)| + |u'|_{L^1(\Omega)}$$

are equivalent.

On the other hand $BV(\Omega)$ is also a Banach space according to the norm

$$|u|_{BV(\Omega)} = |u(0)| + V_0^1(u),$$

as long as all $u \in BV(\Omega)$ have the properties of being continuous to the right in internal points of Ω, and with $V_0^1(u)$ as the total variation of u in Ω.

It is, however, known that every $u \in BV(\Omega)$ identifies, and is identified with the exception of a constant, a measure μ_u completely regular on Ω, and that the set M of all these regular measures can be identified with C^*, the dual of the Banach space $C = C(\Omega,\mathbb{R})$. From this we conclude that $BV(\Omega)$ can be identified with $\{\mathbb{R} \oplus C(\Omega)\}^*$, with the duality between $BV(\Omega)$ and $\{R \oplus C(\Omega)\}$ expressed by

$$a \cdot u(0) + \int_\Omega f \, d\mu_u \text{ for } (a,f) \in \mathbb{R} \oplus C(\Omega),\ u \in BV(\Omega).$$

At this point, for the sake of brevity, let us denote by w^* the weak topology induced on $BV(\Omega)$ by $R \oplus C(\Omega)$, a topology for which $(a_\alpha, u_\alpha) \to (a,u)$ if and only if $a_\alpha \to a$ and $\int_\Omega g\, u_\alpha dt \to \int_\Omega gu\, dt$ for every $g \in C(\Omega)$. It can be noticed that $W^{1,1}(\Omega)$ is dense in $BV(\Omega)$ with respect to w^* and, therefore, having assigned a functional

$$I : u \in W^{1,1}(\Omega) \to I(u) \in [-\infty, +\infty],$$

we can introduce the functional on $BV(\Omega)$:

$$J(u) = \inf \{\underline{\lim}\ I(v_\alpha),\ v_\alpha \in W^{1,1}(\Omega),\ v_\alpha \xrightarrow[w^*]{} u\}$$

with v_α as a generalized (Moore-Smith) sequence which converges to u in the topology w^*.

It is immediate that, if $v_\alpha \xrightarrow[w^*]{} u$, then

$$v_\alpha(0) \to u(0), \quad v_\alpha(1) \to v(1).$$

In particular, if a normal integrand $f : \Omega \times \mathbb{R} \times \mathbb{R} \to [-\infty, +\infty]$ is given, such that f is $\mathcal{L}(\Omega) \otimes B(\mathbb{R} \times \mathbb{R})$-measurable, and $f(t,\cdot,\cdot)$ is lower semicontinuous in $\mathbb{R} \times \mathbb{R}$, and if for every $u \in W^{1,1}(\Omega)$ we take

$$I(u) = \int_\Omega f(t,u(t), u'(t))dt$$

with the usual conventions, then $J : u \in BV(\Omega) \to J(u) \in [-\infty, +\infty]$ is thus defined.

At this point it can be shown (see O. Caligaris, F. Ferro, P. Oliva [1]) that

(1) $J(u) \leqq I(u)$ for every $u \in W^{1,1}(\Omega)$;

(2) $\inf \{J(u) : u \in BV(\Omega)\} = \inf \{I(u) : u \in W^{1,1}(\Omega)\}$;

(3) J is w*-lower semicontinuous on $BV(\Omega)$;

(4) I is w*-lower semicontinuous on $W^{1,1}(\Omega)$ if and only if $I(u) = J(u)$ for every $u \in W^{1,1}(\Omega)$;

(5) J attains its minimum in every w*-closed and bounded set of $BV(\Omega)$;

(6) If there exist $K \in \mathbb{R}_+$, $h \in L^1(\Omega)$, such that

$$f(t,s,v) \geqq K \{|s| + |v|\} - h(t)\},$$

then J has a minimum in $BV(\Omega)$;

(7) Under the assumption that there exist also $A \in \mathbb{R}_+$, $g \in L^1(\Omega)$, such that

$$K|v| - h(t) \leqq f(t,s,v) \leqq A(|s| + |v| + g(t)),$$

then

$$J(u) = I(u) \text{ for every } u \in W^{1,1}(\Omega).$$

16. Another interesting problem is that of the existence of integral representations of $J(u)$ in $BV(\Omega)$. In relation to this problem, Caligaris, Ferro, and Oliva [1] have proved that:

(8) If f is a normal integrand, with $f(t,s,v) = f(t,v)$ and satisfying suitable properties, among them those of linear growth and continuity in v, then for every $u \in BV(\Omega)$ the following holds:

$$J(u) = \int_\Omega f(t,u'(t))dt + \int_\Omega f_r(t,\xi(t),0)d\theta,$$

where

$$f_r(t,z,0) = \lim_{\tau\to 0} f(t, \frac{z}{\tau})\tau$$

(this limit exists due to the above mentioned linearity property of f; where μ_u is the regular measure on Ω associated with u; where $\mu_u = \mu_a + \mu_s$ is the decomposition of μ_u in the absolutely continuous part μ_a, with respect to the Lebesgue measure on Ω, and μ_s is the singular part, and where finally $\xi(t) = d\mu_1/d\theta$, where θ is a measure on Ω absolutely continuous with respect to the Lebesgue measure.

17. Subsequently, all these results have been extended by F. Ferro [5], [6], [7], [8] to the n-dimensional case. The following is just a rapid view on these results.

Let Ω be an open bounded set of $\mathbb{R}^n$, whose boundary $\partial\Omega$ belongs to the class $C^{1,0}$. Let us introduce the class BV of functions $u : \Omega \to \mathbb{R}$:

$$BV(\Omega) = \{u \in L_1(\Omega), \quad \nabla u \in M_n(\Omega) = \{C(\bar{\Omega})^n\}^*,$$

with ∇u is the gradient of u intended in the distributional sense, and $M_n(\Omega)$ the n-dimensional regular measures on Ω whose total variation in Ω is bounded.

If in $BV(\Omega)$ we introduce the norm

$$|u|_{BV(\Omega)} = |u|_{L^1(\Omega)} + |\nabla u|_{M_n(\Omega)},$$

where $|\nabla u|_{M_n(\Omega)}$ is the total variation of ∇u on Ω, then $BV(\Omega)$ becomes a Banach space.

On the other hand, if we consider the Banach space

$$\mathbb{R} \oplus M_n(\Omega) = \{\mathbb{R} \oplus \{C(\bar{\Omega})\}^n\}^*$$

with the norm

$$|(a,u)| = |a| + |v|_{M_n(\Omega)},$$

then, as in the 1-dimensional case, $\mathbb{R} \oplus M_n(\Omega)$ may be provided with the weak topology $w^* = \sigma(\mathbb{R} \oplus M_n, \mathbb{R} \oplus \{C(\bar{\Omega})\}^n)$. But BV can be considered as a subspace of $\mathbb{R} \oplus M_n(\Omega)$ since every $u \in BV(\Omega)$ determines $(\int_\Omega u\, dt, \nabla u)$, and vice versa this couple completely determines $u \in BV(\Omega)$.

It can be shown that

(1) $W^{1,1}(\Omega)$ is w^*-dense in $BV(\Omega)$;

(2) $BV(\Omega)$ is w^*-closed in $\mathbb{R} \oplus M_n(\Omega)$. Furthermore, if $u_\alpha \xrightarrow[w^*]{} u$, then $u_\alpha \xrightarrow[L^1(\Omega)]{} u$;

(3) In $BV(\Omega)$ the norms

$$|u|_{L^1(\Omega)} + |\nabla u|_{M_n(\Omega)}, \int_\Omega u\,dt + |\nabla u|_{M_n(\Omega)}$$

are equivalent:

(4) Every $u \in BV(\Omega)$, for $\Omega \in C^{1,0}$, admits an exterior trace $\gamma^+(u) \in L^1(\Omega)$ such that

$$\int_\Omega G\,\nabla u + \int_\Omega u \operatorname{div} G = \int_{\partial\Omega} \gamma^+(u)\,dH_{n-1}$$

for every $G \in \{C_0^1(\mathbb{R}^n)\}^n$, where H_{n-1} is the Hausdorff measure on $\partial\Omega$.

We refer to Ferro [10] for the definition of the exterior normal to Ω.

18. At this point, just as in the 1-dimensional case, given an integrand $f : \Omega \times \mathbb{R} \times \mathbb{R}^n \to [-\infty, +\infty]$ which is $\mathcal{L}(\Omega) \otimes B(\mathbb{R} \times \mathbb{R}^n)$-measurable, and having defined

$$I(u) = \int_\Omega f[t,u(t), (\nabla u)(t)]dt$$

for every $u \in W^{1,1}(\Omega)$, we can also consider for every $u \in BV(\Omega)$, the corresponding Serrin integral

$$J(u) = \inf\{\underline{\lim}\, I(u_\alpha),\ u_\alpha \in W^{1,1}(\Omega),\ u_\alpha \xrightarrow[w^*]{} u\}$$

Also in this case we obtain the following:

(1) $J(u) \leq I(u)$ for every $u \in W^{1,1}(\Omega)$;

(2) $\operatorname{Inf}\{J(u) : u \in BV(\Omega)\} = \inf\{I(u) : u \in W^{1,1}(\Omega)\}$;

(3) $J(u) = I(u)$ for every $u \in W^{1,1}(\Omega)$ if and only if $I(u)$ is w^* lower semicontinuous in $W^{1,1}(\Omega)$;

(4) $J(u)$ is w^*-lower semicontinuous in $BV(\Omega)$;

(5) J has a minimum on every bounded w^*-closed set of $BV(\Omega)$;

(6) If there exist $k \in \mathbb{R}_+$, $h \in L^1(\Omega)$, such that

$$f(t,s,v) \geq k(|s| + |v|) - h(t),$$

then J attains a minimum on all $BV(\Omega)$.

(7) If, furthermore, there exist $A \in \mathbb{R}_+$, $g \in L^1(\Omega)$, such that

$$k|v| - h(t) \leq f(t,s,v) \leq A(g(t) + |s| + |v|),$$

and if f satisfies a suitable proper continuity assumption in (t,s), then $J(u) = I(u)$ for every $u \in W^{1,1}(\Omega)$.

Finally, there is an integral representation theorem of J(u) under suitable assumptions. A very general result of this type has been obtained by G. Dal Maso [14].

19. Finally, I shall mention further extensions, depending on frontier values, studied by Ferro in [8].

Besides the properties we have already seen on $BV(\Omega)$, let us consider the space $\{C(\partial\Omega)\}^*$ provided with w_1^*-weak topology of the dual space, and let us note that

$$L^1(\partial\Omega) \subset \{C(\partial\Omega)\}^*.$$

Setting for every $\phi \in L^1(\partial\Omega)$ and $g \in C(\partial\Omega)$,

$$\langle\phi,g\rangle = \int_{\partial\Omega} g\phi \, dH_{n-1},$$

let us identify $W^{1,1}(\Omega)$ with the set of pairs

$$(u,\gamma(u)) \in BV(\Omega) \times \{C(\partial\Omega)\}^*.$$

At this point, it can be shown that $W^{1,1}(\Omega)$ is $w^* \otimes w_1^*$ dense in $BV(\Omega) \oplus (C(\partial\Omega))^*$, and I(u) defined on $W^{1,1}(\Omega)$, can be extended, obtaining

$$J_1(u,\psi) = \inf \{\underline{\lim}\, I(u_\alpha),\ (u_\alpha,\gamma(u_\alpha)) \xrightarrow[w^* \otimes w_1^*]{} (u,\psi)\}$$

for every $(u,\psi) \in BV(\Omega) \oplus \{C(\partial\Omega)\}^*$.

Properties similar to those already seen for previous extensions can be stated for $J_1(u,\psi)$. Here is an example.

Let $\psi \in L^1(\partial\Omega)$, $u \in BV(\Omega)$, and for every $v \in W^{1,1}(\Omega)$ let us take

$$I(v) = \int_\Omega \{1 + |(\nabla u)(t)|^2\}^{1/2}\, dt.$$

Then

$$J_1(u,\psi) = \int_\Omega \{1 + |(\nabla u)(t)|^2\}^{1/2} + \int_{\partial\Omega} |\psi - \gamma(u)|\, dH_{n-1}$$

where the first integral is

$$\sup \{ \int_\Omega u \operatorname{div} G + g_{n+1} : G = (g_1,\dots,g_n), \sum_{i=1}^{n+1} g_i^2 \leq 1, g_i \in C_o^1(\Omega)\},$$

in harmony with the theory of area of nonparametric surfaces.

BIBLIOGRAPHY

[VP] V. Volterra and J. Pérès, Théorie Générale des Fonetionnelles. Gauthier-Villars 1936.

[L] J.L. Lions, Contrôle Optimal de Systèmes Gouvernés par des Equations aux Dérivées Partielles, Dunod 1968.

[C] L. Cesari, Optimization - Theory and Applications. Springer-Verlag 1983.

A. Problems of the calculus of variations in Banach spaces

[1] Ada Bottaro Aruffo, Teoremi di esistenza del minimo per problemi di Bolza e di Lagrange in spazi di Banach, Ann. Mat. Pura Appl. (IV) 123 (1980) 105-160.

[2] Ada Bottaro Aruffo, Teoremi di esistenza del minimo per problemi di Bolza con stato a valori in spazi di Banach, Ann. Mat. Pura Appl. (IV) 125 (1980) 367-389.

[3] Ada Bottaro Aruffo, Problemi rilassati del Calcolo delle Variazioni con stato a valori in spazi di Banach, Pubblicazione del Laboratorio per la Matematica Applicata del C.N.R., Sezione Prima, n. 87 (1979).

[4] Ada Bottaro Aruffo, Su alcune proprietà delle topologie sequenziali e delle integrande normali, Ann. Mat. Pura Appl. (IV) 134 (1983) 27-65.

[5] Ada Bottaro Aruffo, Su alcune estensioni del teorema di Scorza Dragoni, Pubblicazione dell'Istituto per la Matematica Applicata del C.N.R., Sezione Prima, n. 165 (1984), accettato per la pubblicazione su "Rendiconti della Accademia Nazionale delle Scienze (detta dei XL) - Parte prima: Memorie di Matematica".

[6] Ada Bottaro Aruffo, $(\mathcal{L} \otimes \mathcal{B}(\rho))$ - misurabilità e convergenza in misura, Pubblicazione dell'Istituto per la Matematica Applicata del C.N.R., Sezione Prima, n. 166 (1984), accettato per la pubblicazione su "Rendiconti dell'Istituto Lombardo - Accademia di Scienze e Lettere - Sez. A: Scienze Matematiche e applicazioni".

[7] Ada Bottaro Aruffo, "Maggiorazioni integrali", Pubblicazione dell'Istituto per la Matematica Applicata del C.N.R., Sezione Prima, n. 170 (1984).

[8] Ada Bottaro Aruffo, Proprietà di inclusione e di uniforme continuità dell'operatore di Nemitskii, Pubblicazione dell'Istituto per la Matematica Applicata del C.N.R., Sezione Prima, n. 167 (1984), inviato per la pubblicazione su "Ricerche di Matematica".

[9] Ada Bottaro Aruffo, Condizioni necessarie e sufficienti per la continuità dell'operatore di sovrapposizione, Pubblicazione dell'Istituto per la Matematica Applicata del C.N.R., Sezione Prima, n. 168, (1984), inviato per la pubblicazione su "Ricerche di Matematica".

[10] Ada Bottaro Aruffo, Sulla rilassazione di alcuni funzionali integrali in dimensione infinita, Pubblicazione dell'Istituto per la Matematica Applicata del C.N.R., Sezione Prima, n. 171 (1984).

[11] G. Bottaro - P. Oppezzi, Semicontinuità inferiore di un funzionale integrale dipendente da funzioni a valori in uno spazio di Banach, Boll. Un. Mat. Ital. (5) 17B (1980) 1290-1307.

[12] G. Bottaro - P. Oppezzi, Condizioni necessarie per la semicontinuità inferiore di un funzionale integrale dipendente da funzioni a valori in uno spazio di Banach, Boll. Un. Mat. Ital. (5) 18B (1981) 47-65.

[13] G. Bottaro - P. Oppezzi, Semicontinuità inferiore di un funzionale integrale dipendente da funzioni di classe L^p a valori in uno spazio di Banach, Annali di Mat. Pura e Appl. (IV) CXXV (1980) 349-365.

[14] G. Bottaro - P. Oppezzi, Rappresentazione con integrali multipli di funzionali dipendenti da funzioni a valori in uno spazio di Banach, a stampa presso l'Istituto per la Matematica Applicata del C.N.R., Sezione Prima, n. 174 (1984) ed accettato per la pubblicazione su Annali di Mat. pura e Appl.

[15] O. Caligaris - P. Oliva, Un teorema di esistenza per problemi di Lagrange in spazi di Banach, Atti Accad. Naz. Lincei Rend. Cl. Sci. Fis. Mat. Natur. 61 (1976), 571-579.

[16] O. Caligaris - P. Oliva, An existence theorm for Lagrange problems in Banach spaces, J. Optimization Theory Appl. 25 (1978), 83-91.

[17] O. Caligaris - P. Oliva, Some remarks about Bolza and Lagrange problems in Hilbert spaces, Boll. Un. Mat. Ital. 15B (1976), 956-967.

[18] O. Caligaris - P. Oliva, Non convex control problems in Banach spaces, Appl. Math. Optim. 5 (1979), 315-329.

[19] O. Caligaris - P. Oliva, A variational approach to existence theorems for time dependent multivalued differential equations, Boll. Un. Mat. Ital. 16B (1979), 612-628.

[20] R. Lucchetti - F. Patrone, Nemitskii's operator and its applications to the lower semicontinuity of integral problems, Indiana University Math. Journal (1980).

[21] P. Oppezzi, Convessità della integranda in un funzionale integrale, Comunicazione XI Congresso U.M.I., Palermo (1979), Sezione 1.

[22] P. Oppezzi, Rappresentazione integrale di funzionali locali in spazi di Sobolev di ordine superiore, in stampa presso l'I.M.A. - C.N.R. Sezione Prima, n. 175 (1984).

[23] J. Balder, On seminormality of integral functionals and their integrands, SIAM Cont. Opt. 1985.

[24] G. Buttazzo - G. Dal Maso, On Nemitskii operators and integral representation of singular integrals, Rend. Mat., 1983.

[25] G. Buttazzo - G. Dal Maso, Integral representation and relaxation of local functionals, Nonlinear Analysis.

[26] G. Buttazzo - G. Dal Maso, A characterization of nonlinear functionals on Sobolev spaces which admit an integral representation with a Carathéodory integrand, Univ. di Udine, 1984.

[27] C. Castaing - P. Clauzure, Semicontinuité des fonctionnelles integrales, Seminaire d'Analyse Convexe Montpellier, 1981.

B. Problems of calculus of variations in classes BV

[1] O. Caligaris - F. Ferro - P. Oliva, Sull'esistenza del minimo per problemi di calcolo delle variazioni relativi ad archi di variazione limitata, Boll. Un. Mat. Ital., (5) 14-B (1977), pp. 340-369.

[2] O. Caligaris - P. Oliva, Problemi di Bolza per archi di variazione limitata ed estensione di funzionali variazionali, Boll. Un. Mat. Ital., (5) 14-B (1977), pp. 772-785.

[3] O. Caligaris - P. Oliva, Sulla caratterizzazione di problemi di calcolo delle variazioni per funzioni di variazione limitata, Boll. Un. Mat. Ital., (5) 15-B (1978), pp. 253-271.

[4] F. Ferro, Sul minimo di funzionali definiti sullo spazio delle funzioni di variazione limitata in n dimensioni, Ann. Mat. Pura Appl., (IV) 117 (1978), pp. 153-171.

[5] F. Ferro, Functionals defined on functions of bounded variation in $\mathbb{R}^n$ and the Lebesgue area, SIAM J. Control and Optimization, vol. 16, N. 5 (1978), pp. 778-789.

[6] F. Ferro, Integral characterization of functionals defined on spaces of BV functions, Rend. Sem. Mat. Univ. Padova, vol. 61 (1979).

[7] F. Ferro, Lower semicontinuity, optimization and regularizing extensions of integral functionals, Pubblicazioni del Laboratorio per la Matematica Applicata del C.N.R.

[8] F. Ferro, Variational functionals defined on spaces of BV functions and their dependence on boundary data, Ann. Mat. Pura Appl..

[9] F. Ferro, A minimal surface theorem with generalized boundary data, Boll. Un. Mat. Ital ., (5) 16-B (1979), pp. 962-971.

[10] F. Ferro, Estensioni di funzionali integrali, Pubbl. I.M.A. (1980).

[11] F. Ferro, Lower semicontinuity of integral functionals and applications, Boll. Un. Mat. Ital., 1982.

[12] F. Ferro, BV spaces on manifolds and functions whose traces are measures, Boll. Un. Mat. Ital. (to appear).

[13] G. Buttazzo - G. Dal Maso, Integral representation on $W^{1,\alpha}(\Omega)$ and $BV(\Omega)$ of limits of variational integrals, Atti Accad. Naz. Lincei Rend. Cl. Sci. Fis. Mat. Natur.

[14] G. Dal Maso, Integral representation of $BV(\Omega)$ of Γ-limits of variational integrals, Preprint Scuola Norm. Sup. Pisa (1979).

[15] M. Giaquinta - G. Modica - J. Soucek, Functionals with linear growth in the calculus of variations, I, II. Comment. Math. Univ. Carolinae, 20 (1979), pp. 143-172.

C. Problems of variational stability

[1] T. Zolezzi, On convergence of minima, Boll. Un. Mat. Ital., 1973.

[2] T. Zolezzi, Characterization of variational perturbations of abstract linear quadratic problems, SIAM J. Control Opt., 1978.

[3] T. Zolezzi, Continuity of generalized gradients and multipliers under perturbations, Math. Op. Res., 1982.

[4] T. Zolezzi, Stability Analysis in optimization, Proceedings Erice, 1984.

[5] R. Lucchetti - F. Mignanego, Variational perturbations of the minimum effort problem, JOTA, 1980.

[6] R. Lucchetti - F. Mignanego, Continuous dependence on the data in abstract problems, JOTA.

D. Well posed problems

[1] T. Zolezzi, On equiwell-set minimum problems, Appl. Math. Optim., 1978.

[2] T. Zolezzi, A characterization of well posed optimal control systems, SIAM J. Control Opt., 1981.

[3] R. Lucchetti, Hadamard and Tyhonov well posedness in optimal control, Math. of Oper. Res., 1982.

[4] R. Lucchetti, Some aspects of the connections between Hadamard and Tyhonov well posedness of convex programs, Boll. Un. Mat. Ital., 1982.

[5] R. Lucchetti - F. Patrone, Hadamard and Tyhonov well posedness of a certain class of convex functions, Journal of Math. Anal. and Appl., 1982.

E. Pareto Problems

[1] O. Caligaris - P. Oliva, Necessary and sufficient conditions for Pareto problems, Boll. Un. Mat. Ital. 18 B (1981), 177-216.

[2] O. Caligaris - P. Oliva, Optimality in Pareto problems, Generalized Lagrangians in Systems and Economic Theory, Laxenourg, dicembre 1979.

[3] P. Oliva, Sulla compattezza degli insiemi di livello nel calcolo delle variazioni, Boll. U.M.I. Analisi Funz. Appl. serie V 18 (1981), 89-105.

[4] O. Caligaris - P. Oliva, Semicontinuità di funzioni vettoriali ed esistenza del minimo per problemi di Pareto, Boll. U.M.I., Analisi Funz. Appl. serie VI 2 (1983), 87-103.

[5] O. Caligaris - P. Oliva, Constrained optimization of infinite dimensional vector valued functions with application to infinite horizon integrals, Pubbl. n. 177 sez. 1 dell'I.M.A. di Genova.

[6] O. Caligaris - P. Oliva, Necessary conditions for local Pareto minima, Pubbl. n. 176 sez. 1 dell'I.M.A. di Genova.

[7] R. Peirone, Γ-limiti e minimi di Pareto, Rend. Accad. Lincei, 1983.

J.P. Cecconi
Istituto di Matematica
Applicata
Università di Genova
16132 Genova
Italy.

S CINQUINI

Tonelli's constructive method in the study of partial differential equations

PART 1. INTRODUCTION

1. On January 15, 1928, L. Tonelli made a communication [45][(*)] to the "R. Accademia delle Scienze dell'Istituto di Bologna" concerning the non-linear integral equation

$$\phi(x) = f(x) + \int_a^x K(x,y,\phi(y))dy.$$

In his communication Tonelli announced, as a future development, a constructive approach toward the existence of at least one solution for the more general functional equation

$$\phi(x) = f(x) + A[x, \overset{x}{\underset{0}{\Phi(y)}}], \qquad (0 \leqq x \leqq 1), \tag{1}$$

where, for every x, A[...] denotes a real number which depends on x and on the values of the unknown function ϕ in the interval (0,x).

In his now famous memoir [46] of the same year, in the Bull. Calcutta Math. Soc., Tonelli followed up his announcement by defining the approximations $\Phi_n(x)$, n = 1,2,..., to problem (1),

$$\begin{aligned} \Phi_n(x) &= f(x), && \text{for } 0 \leqq x \leqq \frac{1}{n}, \\ \Phi_n(x) &= f(x) + A[x, \Phi_n(\overset{x-1/n}{\underset{0}{y}})], && \text{for } \frac{1}{n} < x. \end{aligned} \tag{2}$$

It is clear that this is a method of approximations (*not* of successive approximations in the usual terminology), in which in each interval $r/n < x \leqq (r+1)/n$, $(r = 1,\ldots,n-1)$, $\Phi_n(x)$ is defined by means of the values already assigned to it in $0 \leqq x \leqq r/n$: briefly but expressively we may say that this is a "back-step process" of approximation.

(*) The numbers in brackets refer to the bibliography at the end of this paper. For further bibliographical information on partial differential equations refer to the monograph [26] by M. Cinquini Cibrario and S. Cinquini, and the recent expositions of S. Cinquini [19] and of M. Cinquini Cibrario [25].

Under ample assumptions, Tonelli proved the existence of a (continuous) solution of (1). Moreover, after having shown by an example, that under the assumptions of the existence theorem, equation (1) may have more than one solution, Tonelli established a uniqueness theorem.

It remained to prove that the solution of (1) depends continuously on the function f(x) and on the functional A[...], and this was proved by S. Cinquini in a paper [15] published a few years later.

2. (a) It was Dario Graffi [32] who realized in 1931 the power of Tonelli's approach, and applied it to a functional Volterra-type equation which extends the vectorial equation

$$\underline{\underline{I}}(t) = f(\underline{\underline{H}}(t) + \alpha\underline{\underline{I}}(t)) + F[t,\underline{\underline{H}}(\tau) \overset{t}{\underset{o}{+}} \alpha\,\underline{\underline{I}}(\tau)].$$

The latter had occurred to Graffi in the study of a hereditary problem of magnetic and dielectrical induction. Here $\underline{\underline{I}}(t)$ is the unknown vector function, and α is a constant homography.

(b) In 1940 R. Nardini [37] extended Tonelli's approach to the functional equation

$$\Psi(x,y) = f(x,y) + \lambda\int_0^x K(x,y,z)\Psi(x,z)dz + A[x,y,\Psi(\overset{x}{\underset{o}{t}}, \overset{b}{\underset{o}{z}})].$$

3. In particular, the existence theorem for the solution of the Cauchy problem for a system of ordinary differential equations can be established by Tonelli's method, as Sansone has done in his monograph [40] in 1941.

4. Again in connection with ordinary differential equations E. Magenes [35] and G. Stampacchia [43], [44] applied Tonelli's method to boundary value problems. Namely Stampacchia in [43] discussed an extension of the Nicoletti problem by proving the existence of at least one solution of the system

$$y(x) = y(a) + \int_a^x f(t,y(t),z(t))dt,$$

$$z(x) = z(a) + \int_a^x g(t,y(t),z(t))dt,$$

which touches two given curves lying on the planes $x = a_1$ and $x = a_2$. In [44] Stampacchia extended this result to the case of n equations. Magenes in [35] proved two theorems extending Satô's previous result on the existence of at least one solution to the equation

$$y'(x) = y'(x_0) + \int_{x_0}^{x} f(t,y(t),y'(t))dt$$

passing through a given point (x_0,y_0) and tangent at least at one point to a given curve $y = g(x)$.

PART 2. PARTIAL DIFFERENTIAL EQUATIONS

5. Much wider and more relevant was the impact of Tonelli's method in the theory of partial differential equations. This endeavour, in which much use was made of the classical results of Giulio Ascoli and C. Arzelà concerning sequences of equibounded and equicontinuous functions, is connected with an attempt by Arzelà in 1906 to prove the existence of at least one solution of the equation

$$\frac{\partial z}{\partial x} = f(x,y,z,\frac{\partial z}{\partial y}) \qquad (3)$$

satisfying the Cauchy condition

$$z(0,y) = \phi(y). \qquad (4)$$

The same endeavour is also connected with an idea of C. Severini (1938) who had thought that the same constructive process, used by Cinquini around 1930 for the semicontinuity of double integrals in the calculus of variations, could also be used toward a solution of problem (3) - (4).

6. It was not until 1947 that a memoir of E. Baiada [3] was concerned with the problem (3) - (4). Let us consider the strip $\Sigma : 0 \leq x \leq a, -\infty < y < +\infty$, and for any integer n let us divide the interval $0 \leq x \leq a$ into n equal parts of length $\delta_n = a/n$. Then Baiada considers the function $z = z_n(x,y)$ defined in Σ by taking

$$z_n(x,y) = \frac{1}{2}[\phi(y + \frac{x}{2}) + \phi(y - \frac{x}{2})] + \int_{y-x/2}^{y+x/2} f(0,v,\phi(v),\phi'(v))dv$$

for $0 \leqq x \leqq \delta_n$;

$$z_n(x,y) = \frac{1}{2}[z_n(r\delta_n, y + \frac{x-r\delta_n}{2}) + z_n(r\delta_n, y - \frac{x-r\delta_n}{2})]$$

$$+ \int_{y-(x-r\delta_n)/2}^{y+(x-r\delta_n)/2} f(r\delta_n, v, z_n(r\delta_n,v), \frac{\partial z_n(r\delta_n,v)}{\partial v})dv$$

for

$$r\delta_n < x \leqq (r+1)\delta_n. \qquad (r = 1,\dots,n-1). \qquad (5)$$

This process is well in the spirit of Tonelli, since the definition of $z_n(x,y)$ in each strip (5) depends on the values of the same function in the preceding r strips. As the original Tonelli approach, it is a process of approximations, *not* of successive approximations in the usual terminology. Initially, Baiada [3] proved the existence of at least one solution z(x,y) (continuous together with its first order partial derivatives) of the Cauchy problem (3) - (4) by the use of an Arzelà's lemma on sequences of equibounded and "equioscillating" functions.

Baiada resumed his own process in successive papers, some in collaboration with C. Vinti, and in 1955 Baiada and Vinti [5] proved the existence of at least one Lipschitz(†) solution of problem (3) - (4). As is well known, the difficulties presented by this problem, in comparison with both the analogous one dimension Cauchy problem

$$y' = f(x,y), \quad y(x_0) = y_0, \qquad (6)$$

and the Darboux problem (8) - (9) below (cf. no. 8), lie in the fact that in equation (3) there appears more than one differential element of maximal order.

On the other hand, in 1953 Baiada [4] had considered equations of the type

$$\frac{\partial z}{\partial x} = f(x, \frac{\partial z}{\partial y}) \qquad (3')$$

(†) In the present exposition we refer mostly to solutions almost everywhere.

in order to clarify which of the hypotheses on the function $f(x,y,z,q)$ in (3) are due to the fact that f depends on $\partial z/\partial y$. For equation (3') Baiada obtained an existence theorem for at least one Lipschitz solution of the Cauchy problem under conditions weak enough so as not to imply uniqueness, as shown by a 1951 example by M. Pagni [38].

7. (a) In 1963 C. Vinti [48], with suitable modifications of Baiada's process, extended Baiada's result to systems of the form

$$\frac{\partial z_i}{\partial x} = f_i(x,y,z_1,\dots,z_n;\ \frac{\partial z_i}{\partial y}), \qquad (i = 1,\dots,n), \tag{7}$$

namely, the existence of a Lipschitz solution but not its uniqueness. This result, and the one of Baiada for problem (3') are most relevant, though it must be said that in both the functions $\phi_i(y)$, $(i = 1,\dots,m)$, defining the initial data, are required to be continuous with continuous derivatives $\phi_i'(y)$, and satisfying inequalities of the type

$$\left|\frac{\phi_i'(y+h)-\phi_i'(y-h)}{2h}\right| < M(y),$$

with $M(y)$ integrable(ǂ).

(b) It should be noted here, that R. Conti [27] in 1952 had, by the use of Baiada's process, already ascertained for system (7) the existence of at least one solution continuous with its first order partial derivatives.

(c) Recently, Baiada's process has been applied to functional equations which contain as a particular case the partial differential equations above. This has been done by P. Brandi [6ii], by P. Brandi and R. Ceppitelli [6iii], and by A. Salvadori [39i]. The latter has considered the system

$$\frac{\partial z^{[i]}}{\partial x} = f^{[i]}(x,y;\ T_1^{[i]}(x, y, z(x,y)),\dots,T_{r_i}^{[i]}(\dots);\ \frac{\partial z^{[i]}}{\partial y}), \qquad (0 \leqq x \leqq a,\ -\infty < y < +\infty),$$

$$z^{[i]}(x,y) = \phi^{[i]}(x,y), \qquad (a_0 \leqq x \leqq 0, -\infty < y < +\infty),$$

where $i = 1,\dots,k$.

(ǂ) In the present exposition, integrability is understood in the sense of Lebesgue.

PART 3. THE DARBOUX PROBLEM

8. (a) The process proposed by Tonelli has been widely applied, with suitable modifications, to second order semilinear equations

$$\frac{\partial^2 z}{\partial x \partial y} = f(x,y,\ z(x,y),\ \frac{\partial z}{\partial x},\ \frac{\partial z}{\partial y}), \qquad (8)$$

particularly in connection with Darboux's problem

$$z(x,0) = \sigma(x), \quad z(0,y) = \tau(y), \qquad \sigma(0) = \tau(0) = z_0. \qquad (9)$$

This problem, that is, the passage from one to two independent variables, constitutes the most natural extension of the Cauchy problem (6). The results have been obtained under classical conditions for the function f(...), as well as under Caratheodory conditions.

(b) First Conti [28] (1953) formulated problem (8) - (9) in the form

$$z(x,y) = \sigma(x) + \tau(y) - z_0 + \int_0^x \int_0^y f(u,v,z(u,v),\ \frac{\partial z}{\partial u},\ \frac{\partial z}{\partial v}) du\, dv, \qquad (10)$$

and defined the sequence of approximations $z^{[n]}(x,y)$, $(n = 1,2,\ldots)$ by considering, for any integer n, the subdivision of the rectangle $R = [0 \leqq x \leqq a,\ 0 \leqq y \leqq b]$ in n parts $R_j^{[n]}$, $(j = 1,\ldots,n)$, by means of the n-1 pairs of segments

$$x = \frac{r}{n} a, \quad \frac{r}{n} b \leqq y \leqq b; \quad \frac{r}{n} a \leqq x \leqq a, \quad y = \frac{r}{n} b, \quad (r = 1,\ldots,n-1).$$

Then Conti defined the approximations by taking

$$z^{[n]}(x,y) = \sigma(x) + \tau(y) - z_0, \quad \text{for } (x,y) \in R_1^{[n]},$$

and then for $(x,y) \in R - R_1^{[n]}$ by taking

$$z^{[n]}(x,y) = \sigma(x) + \tau(y) - z_0$$

$$+ \int_0^{x-a/n} \int_0^{y-b/n} f(u,v,\ z^{[n]}(u,v),\ \frac{\partial z^{[n]}}{\partial u},\ \frac{\partial z^{[n]}}{\partial v}) du\, dv.$$

For $\sigma(x)$, $\tau(y)$ both Lipschitz, Conti [28] proved the existence of at least one Lipschitz solution z for equation (10).

Later, in 1958 Conti [29] extended his result by assuming $\sigma(x)$ and $\tau(y)$ both absolutely continuous, and proving the existence of at least one solution of (10) in the class Γ_{12}. Here Γ_{12} is the class of all functions $z(x,y)$ which are doubly absolutely continuous (that is absolutely continuous in the sense of Vitali), and which are absolutely continuous as functions of x only for all y, and of y only for all x. To prove this result, Conti considered the expressions

$$P(x,y) = \sigma(x) - z_0 + \int_0^x \int_0^y f(u,v,\tau(v) + P(u,v), \frac{\partial P}{\partial u}, \frac{\partial Q}{\partial v}) du\, dv,$$

$$Q(x,y) = \tau(y) - z_0 + \int_0^x \int_0^y f(u,v,\tau(v) + P(u,v), \frac{\partial P}{\partial u}, \frac{\partial Q}{\partial v}) du\, dv,$$

and first remarked that

$$z(x,y) = \tau(y) + P(x,y) = \sigma(x) + Q(x,y)$$

is a solution of (10). Then Conti defined two sequences of functions which approximate $P(x,y)$, $Q(x,y)$ respectively, by taking

$$P_n(x,y) = \sigma(x) - z_0 \qquad \text{for } 0 \leqq x \leqq a, \quad 0 \leqq y \leqq \frac{b}{n};$$

$$Q_n(x,y) = \tau(y) - z_0 \qquad \text{for } 0 \leqq x \leqq \frac{a}{n}, \quad 0 \leqq y \leqq b;$$

$$P_n(x,y) = \sigma(x) - z_0 + \int_0^x \int_0^{y-b/n} f(u,v,\tau(v) + P_n(u,v), \frac{\partial P_n}{\partial u}, \frac{\partial Q_n}{\partial v}) du\, dv$$
$$\text{for } 0 \leqq x \leqq a, \quad \frac{b}{n} < y \leqq b;$$

$$Q_n(x,y) = \tau(y) - z_0 + \int_0^{x-a/n} \int_0^y f(u,v,\tau(v) + P_n(u,v), \frac{\partial P_n}{\partial u}, \frac{\partial Q_n}{\partial v}) du\, dv$$
$$\text{for } \frac{a}{n} < x \leqq a, \quad 0 \leqq y \leqq b.$$

In other words, the back step process is used only either on the variable x, or on the variable y.

Two papers on the subject by A. Zitarosa [51] and by F. Guglielmino [34] appeared in the years 1959-60. Zitarosa proved an existence theorem for at least one solution of the problem (8) - (9) in the class C_{12} of the functions $z(x,y)$ continuous together with their partial derivatives $\partial z/\partial x$, $\partial z/\partial y$, $\partial^2 z/\partial x \partial y$, by using methods of functional analysis. However, his work bears on Tonelli's approach, since Zitarosa has pointed out that the continuity of the basic transformation underlying his research, can be proved by the same argument of Cinquini in paper [15], as mentioned in no. 1 and independently from fixed point theorems. Guglielmino has also proved the result of Zitarosa by using Tonelli's approach, namely, by means of a suitable adaptation of the approximations presented by Conti in [29]. In connection with the work of Guglielmino on equation (8), the remark should be made that he sometimes makes use of the uniqueness of the solution to the Cauchy problem (6), and in particular of the 1925 Tonelli theorem. On the other hand, in Guglielmino's result is contained the 1959 work of W. Walter, who, for any $\alpha > 0$, had constructed the functions

$$z_\alpha(x,y) = \sigma(x) + \tau(y) - z_0 + \int_0^x \int_0^y f(u,v,z_\alpha(u-\alpha,v),p_\alpha(u,v-\alpha),q_\alpha(u-\alpha,v))dudv,$$

$$p_\alpha(x,y) = \sigma'(x) + \int_0^y f(x,v,z_\alpha(x-\alpha,v),\ p_\alpha(x,v-\alpha),\ q_\alpha(x-\alpha,v))dv,$$

$$q_\alpha(x,y) = \tau'(y) + \int_0^x f(u,y,z_\alpha(u-\alpha,y),\ p_\alpha(u,y-\alpha),\ q_\alpha(u-\alpha,y))du.$$

Here Walter uses the "back step process" in a way that is very different from what had been done before. This having been mentioned, Walter considered sequences $\alpha = \alpha_n$, and took limits as $\alpha_n \to 0+$.

(c) Again in connection with Tonelli's backstep process, but under Carathéodory's conditions for the function f(...), mention should be made of a 1958 note of Guglielmino [33], and a paper by G. Pulvirenti [39] concerning first of all the particular case of the semilinear equations

$$z(x,y) = \phi(x,y) + \int_0^x \int_0^y f(u,v,z(u,v))du\,dv \qquad (11)$$

with

$$\phi(x,y) = \sigma(x) + \tau(y) - z_0.$$

In [39] G. Pulvirenti studies the Peano phenomenon for equation (11), and in this paper, though making use of methods of functional analysis, considered a sequence of functions $v_n = v_n(x,y)$, $(n = 1,2,\ldots)$, with

$$v_n(x,y) = z(x,y) - \phi(x,y), \qquad \text{for } (x,y) \in R_1^{[n]},$$

$$v_n(x,y) = z(x,y) - \phi(x,y) - \int_0^{x-a/n}\int_0^{y-b/n} f(u,v,z(u,v))du\, dv,$$
$$\text{for } (x,y) \in R-R_1^{[n]},$$

where the subdivision of the rectangle R in the parts $R_j^{[n]}$ is the same as that already mentioned in connection with the paper [28] by Conti.

With regard to equation (10), mention should be made of the conclusive result of G. Arnese [2], who proved in 1963 that, if $\sigma(x)$, $\tau(y)$ are absolutely continuous, then the existence of at least one solution z(x,y) belonging to the class Γ_{12} (already considered by Conti in [29]) can be guaranteed under rather wide conditions on the function f(...).

(d) The Tonelli method has been applied also to the Picard problem concerning the existence in $\Delta = \Delta_x \times \Delta_y$ ($\Delta_x = (-c,c')$, $\Delta_y = (-d,d')$, $c,c', d, d' \geqq 0$, $c + c' > 0$, $d + d' > 0$) of a solution of equation (8) satisfying

$$z(x,0) = \mu(x),\ x \in \Delta_x, \quad z(\beta(y),y) = \nu(y),\ y \in \Delta_y,$$

where $\mu(x)$ is a given function in Δ_x, $\beta(y)$, $\nu(y)$ are given functions in Δ_y, with $\mu(0) = \nu(0)$, $\beta(0) = 0$, $-c \leqq \beta(y) \leqq c'$. In 1959 in the paper [34] mentioned above, Guglielmino proved the existence of at least one solution to equation (8) in the Class C_{12}. Later, in 1967, in a relevant memoir G. Santagati [42], not only proved an existence theorem in the class Γ_{12}, but also pointed out that many results already proved by others for the Darboux problem can be derived as particular cases. On the other hand, N. Fedele [30], [31], by continuing Arnese's analysis [1], [2], proved two theorems of approximation for the Picard problem in the classes C_{12} and Γ_{12} respectively.

This last result is in harmony with the remark that Tonelli's method can be used to approximate any solution of equation (1) of which the existence has already been proved.

9. (a) For lack of space we cannot present here the application of Tonelli's method to the Darboux problem for systems of equations, and the results obtained between 1958 and 1966 by G. Villari [47], by L. Merli [36], and by N. Fedele in the papers [30], [31] mentioned above.

M.G. Cazzani Nieri [9] considered systems of the form

$$\sum_{j=1}^{m} a_{ij}(x,y,z_1,\dots,z_m)\frac{\partial z_j}{\partial x} = f_i(x,y, z_1,\dots,z_m), \quad (i = 1,\dots,r),$$

$$\sum_{j=1}^{m} a_{ij}(x,y,z_1,\dots,z_m)\frac{\partial z_j}{\partial y} = f_i(x,y, z_1,\dots,z_m), \ (i = r+1,\dots,m),$$

(where $1 \leqq r < m$), and proved the existence of at least one Lipschitz solution, satisfying boundary conditions also defined by Lipschitz functions.

Again for lack of space we refer to the extensive and relevant 1965 memoir of G. Santagati [41], in which the author, for systems of two equations, obtained a set of results of which we mention only the following ones.

If we take

$$\sigma(x) = z_0 + \int_0^x \phi(t)dt, \quad \tau(y) = z_0 + \int_0^y \psi(t)dt, \quad (\sigma(0) = \tau(0) = z_0),$$

then, for the problem

$$\begin{cases} \dfrac{\partial z_1}{\partial x} = F(x,y, \sigma(x) + \displaystyle\int_0^y z_1(x,t)dt, z_1(x,y), z_2(x,y)), \\ \dfrac{\partial z_2}{\partial y} = G(x,y, \tau(y) + \displaystyle\int_0^x z_2(t,y)dt, z_1(x,y), z_2(x,y)), \end{cases} \tag{12}$$

$$z_1(0,y) = \psi(y), \quad z_2(x,0) = \phi(x), \tag{13}$$

Santagati proves the existence of a solution $z_1(x,y)$, $z_2(x,y)$, of which $z_1(x,y)$ is absolutely continuous in x (for all y) and integrable in y (for all x), and analogously for $z_2(x,y)$ (by exchanging x and y).

Santagati considers also the problem analogous to the one above, under the same boundary data (13), and for which, in the second equation (12), $\tau(y) + \int_0^x z_2(t,y)dt$ is replaced by $\sigma(x) + \int_0^y z_1(x,t)dt$.

(b) Concerning the problem

$$\begin{cases} \dfrac{\partial^3 w}{\partial x \partial y \partial z} = f(x,y,z,\ w(x,y,z),\ \dfrac{\partial w}{\partial x}, \dfrac{\partial w}{\partial y}, \dfrac{\partial w}{\partial z}) \\ w(x,y,0) = w(x,0,z) = w(0,y,z) = 0, \end{cases}$$

L. Castellano [8] in his memoir of 1968, by using Tonelli's approach, not only reproved the existence theorem that M. Frasca had obtained in 1966 by methods of functional analysis, but also obtained a theorem of approximation of any solution to the given problem.

PART 4. SYSTEMS OF HYPERBOLIC EQUATIONS

10. (a) In this last Section we present results obtained by Tonelli's method (as well as later results obtained by different methods) for systems of quasi linear partial differential equations of the hyperbolic type in the bicharacteristic form, namely for the case of $r + 1$ independent variables, $r > 1$(§), of the form

$$\sum_{j=1}^{m} a_{ij}(x,y_1,\dots,y_r;\ z_1,\dots,z_m)\Big[\frac{\partial z_j}{\partial x} + \sum_{k=1}^{r} \lambda_{ik}(x,y_1,\dots,y_r;\ z_1,\dots,z_m)\frac{\partial z_j}{\partial y_k}\Big] = f_i(\dots),\ (i = 1,\dots,m), \tag{14}$$

where $x,y_1,\dots,y_r$ are the independent variables, $z_j(x,y_1,\dots,y_r)$, $(j=1,\dots,m)$, are the unknown functions, and $f_i(\dots) = f_i(x,y_1,\dots,y_r;\ z_1,\dots,z_m)$ are given functions of $x,y_1,\dots,y_r,\ z_1,\dots,z_m$. It is well evident that in this research, one has encountered the same basic difficulty mentioned in no. 6 for a single equation.

(§) Concerning the reduction of a general quasi-linear hyperbolic system in $r+1$ independent variables, $(r > 1)$, to the bicharacteristic form (14), let us mention that S. Cinquini [19], by continuing his work from 1967, proved that the necessary and sufficient condition in order that such a reduction is possible, is the permutability of certain r matrices. Moreover, such permutability has a relevance in applications (see Cinquini [20]). P. Bassanini pointed out ([6], 1982), as stated by P. Bouillat in 1965, that such permutability is a necessary and sufficient condition in order that in wave propagation there are no linear shocks.

(b) In 1955 the initial work of M. Cinquini Cibrario [21] concerned systems with only two independent variables of the form

$$\frac{\partial z_i}{\partial x} + \lambda_i(x,y;z_1,\ldots,z_m)\,\frac{\partial z_i}{\partial y} = f_i(x,y;z_1,\ldots,z_m), \quad (i = 1,\ldots,m). \quad (15)$$

Then, in 1957 M. Cinquini Cibrario [22] considered systems with any number r+1 of independent variables of the form

$$\frac{\partial z_i}{\partial x} + \sum_{k=1}^{r} \lambda_{ik}(x,y_1,\ldots,y_r;\ z_1,\ldots,z_m)\,\frac{\partial z_i}{\partial y_k} = f_i(\ldots), \quad (i=1,\ldots,m), \quad (16)$$

which S. Yosida has called "the S. Cinquini equations".

In 1965 Cinquini Cibrario [23] considered another case of systems (14), namely the semilinear systems in the bicharacteristic form

$$\sum_{j=1}^{m} a_{ij}(x,y_1,\ldots,y_r)\left[\frac{\partial z_j}{\partial x} + \sum_{k=1}^{r} \lambda_{ik}(x,y_1,\ldots,y_r)\,\frac{\partial z_j}{\partial y_k}\right] = f_i(\ldots), \quad (i=1,\ldots,m), \quad (17)$$

and in both systems (16) and (17) the second members are the same as in system (14).

For all systems (14) to (17) the author considered the Cauchy problem

$$z_i(0,y_1,\ldots,y_r) = \phi_i(y_1,\ldots,y_r), \qquad (i = 1,\ldots,m), \quad (18)$$

where the functions $\phi_i(y_1,\ldots,y_r)$ are assumed to be Lipschitzian in the complex of the variables $y_1,\ldots,y_r$ in the rectangle $R = [-b_k \leq y_k \leq b_k;\ k=1,\ldots,r]$, while the solutions of the Cauchy problem are sought in the class G of the functions $z_j(x,y_1,\ldots,y_r)$, $(j=1,\ldots,m)$, each z_j being absolutely continuous as a function of x alone (for every fixed r-pla $(y_1,\ldots,y_r)$), and Lipschitzian in the complex of the variables $y_1,\ldots,y_r$ with the Lipschitz constant independent of x. Of course, as mentioned already, we are dealing here with solutions in the sense that the system of equations is required to be satisfied almost everywhere, while relations (18) are satisfied everywhere in R(¶)

(¶) Let us state here explicitly that, in the next lines, when we refer to systems (14)and (17) and initial data (18) in the case of only two independent variables, we shall use the notations (14*), (17*), (18*) without rewriting the systems (cf. no. 12(b), no. 13).

The functions $\lambda_{ik}(\ldots)$, $f_i(\ldots)$ are assumed to satisfy Carathéodory type conditions of which, for the sake of brevity, we mention here only the following one: there are r non-negative, quasi-continuous, integrable functions $L_k(x)$, $k = 1,\ldots,r$, such that

$$|\lambda_{ik}(x,y_1,\ldots,y_r)| \leq L_k(x), \qquad (i=1,\ldots,m;\ k=1,\ldots,r);$$

while the functions $a_{ij}(\ldots)$, $(i,j=1,\ldots,m)$ are assumed to satisfy classical conditions with

$$A = \det (a_{ij}) = 1. \tag{19}$$

The existence of at least one solution of the Cauchy problem is established in a region

$$T_r:\ 0 \leqq x \leqq a, \quad -b_k + \int_0^x L_k(t)dt \leqq y_k \leqq b_k - \int_0^x L_k(t)dt, \ (k=1,\ldots,r),$$

where a is a positive number such that for every $x < a$ we have

$$\int_0^x L_k(t)dt < b_k, \qquad (k=1,\ldots,r), \tag{20}$$

where for $x = a$ the equality sign may hold.

(c) In papers [21], [22], [23] Tonelli's method has been applied with suitable devices to systems (15), (16), (17).

For the sake of brevity we refer only to system (17), and we mention, first of all, that the "back step process" is applied, not to system (17), but to the system that Cinquini Cibrario obtains from (17), namely $(j=1,\ldots,m)$

$$\begin{aligned} z_j(x,y_1,\ldots,y_r) &= \Psi_j(x,y_1,\ldots,y_r) \\ &+ \sum_{i=1}^{m} A_{ij}(x,y_1,\ldots,y_r) \int_0^x \Big[\sum_{s=1}^{m} \frac{da_{is}(X,g_{i1},\ldots,g_{ir})}{dX} z_s(X,g_{i1},\ldots,g_{ir}) \\ &+ f_i(X,g_{i1},\ldots,g_{ir};\ z_1(X,g_{i1},\ldots,g_{ir}),\ldots,z_m(\ldots))\Big]dX, \end{aligned} \tag{21}$$

where A_{ij} is the algebraic complement of a_{ij} in the determinant A. Also, in (21) for every $i = 1,\ldots,m$ and by force of Carathéodory's theorem,

$g_{ik} = g_{ik}(X;x,y_1,\dots,y_r)$ $(k = 1,\dots,r)$ is the unique solution of the system of ordinary differential equations

$$g_{ik} = y_k - \int_X^X \lambda_{ik}(t,g_{i1},\dots,g_{ir})dt, \quad (k = 1,\dots,r).$$

Finally, in (21) the functions $\Psi_j(\dots)$ are so defined (cf. [23]) that

$$\Psi_j(0,y_1,\dots,y_r) = \phi_j(y_1,\dots,y_r).$$

For any integer $n \geqq 2$, the interval $(0,a)$ is divided into n equal parts of equal length $\delta_n = a/n$, and Cinquini Cibrario defines the m sequences of functions (obtained by letting j run over $j = 1,\dots,m$), say $z_j^{[n]}(x,y_1,\dots,y_r)$, $n = 2,3,\dots$, by taking for $x \leqq 0$

$$z_j^{[n]}(x,y_1,\dots,y_r) = \phi_j(y_1,\dots,y_r),$$

and then in T_r

$$z_j^{[n]}(x,y_1,\dots,y_r) = \Psi_j(x,y_1,\dots,y_r)$$

$$+ \sum_{i=1}^{m} A_{ij}(\dots) \int_0^X [\sum_{s=1}^{m} \frac{da_{is}(\dots)}{dX} z_s^{[n]}(X - \delta_n, g_{i1},\dots,g_{ir})$$

$$+ f_i(X,g_{i1},\dots,g_{ir};z_1^{[n]}(X - \delta_n, g_{i1},\dots,g_{ir}),\dots,z_m^{[n]}(\dots))]dX,$$

where $A_{ij}(\dots)$, $da_{is}(\dots)/dX$ do not differ from the analogous expressions in (21). In other words, the "backstep process" is used here in a way which differs from that mentioned in the previous paragraph, namely the process is applied only to the variable X of the functions $z_j^{[n]}(X,g_{i1},\dots,g_{ir})$.

(d) Finally, by the use of the form mentioned above of Tonelli's back step process for the semilinear system (17), and under the conditions stated in (b), Cinquini Cibrario [24] has proved the existence of at least one solution to problem (14) - (18) by using the method of successive approximations.

(e) The existence theorem of at least one solution of the Cauchy problem for the quasi linear system (14) has been investigated by L. Cesari ([14i], 1974), who gave a new proof of the result, under the same assumptions as in paper [24] by Cinquini Cibrario, by the use of Banach's fixed point theorem, from which also properties of uniqueness follow.

In the same paper [14i], Cesari considered also the boundary value problem defined by the requirements

$$\sum_{j=1}^{m} b_{ij}(y_1,\dots,y_r)z_j(a^{[i]},y_1,\dots,y_r) = \Psi_i(y_1,\dots,y_r), \qquad (i=1,\dots,m), \quad (18')$$

where $a^{[i]}$, $(i = 1,\dots,m)$ are numbers satisfying the relation $0 \leqq a^{[i]} \leqq a$. Cesari proved an existence theorem of at least one solution to problem (14), (18'), under the assumption that the matrices (a_{ij}), (b_{ij}) have "dominant main diagonal" in a rather strong sense. For the lack of space, we refer to [14i] for the precise definition.

(f) Cesari applied the existence theorem, mentioned in (e) for the boundary value problem, to a case of resonance in non-linear optics, mentioned by Graffi [32i] in 1967, which concerns a phenomenon observed experimentally when a monochromatic laser beam crosses a thin crystal. This application was resumed and continued in papers by Bassanini and Cesari. We mention here only paper [6i] (1980), for a method of successive approximations.

11. (a) First, we have to add that in the papers [21], [22], [23], [24] mentioned in no. 10, Cinquini Cibrario also proved that, under the same conditions used for existence, the solution is unique and depends continuously on the data.

(b) On the other hand, in 1955 for systems (15) and then in 1967 for systems (14), S. Cinquini [17], [18] proved uniqueness theorems for the solutions of the Cauchy problem under weaker assumptions, both on the initial data since the functions $\phi_j(y_1,\dots,y_r)$ are assumed only to be continuous, and for that which concerns the class of functions in which the solution is sought, since each function $z_j(x,y_1,\dots,y_r)$ is assumed to be globally continuous, absolutely continuous as a function of x only for almost all the r-mples $(y_1,\dots,y_r)$, and absolutely continuous as a function of each y_k for almost all x and for all (r-1)-mples $(y_1,\dots,y_{k-1},y_{k+1},\dots,y_r)$. The functions z_j are also assumed to satisfy, for almost all x, an inequality, for which we refer to paper [18], of the type

$$|z_j(x,y_1,\dots,y_r) - z_j(x,\bar{y}_1,\dots,\bar{y}_r)| \leqq \sum_{k=1}^{r} \omega_{jk}(x)|y_k - \bar{y}_k|.$$

Cinquini obtained this greater generality by making use of an approximation scheme by integral means which he had already used in other chapters of

Mathematical Analysis, and which had its origins in Tonelli's work of 1928 on the functions of two variables of bounded variation.

(c) It is of some interest to point out here that the existence and uniqueness theorems on the Cauchy problem are such that, if we apply any one of them starting from $x = 0$ (that is, in any region T_r (no. 10(b))), then we can apply the same theorem starting from $x = a'$ with $0 < a' < a$.

12. (a) Before going into more recent work on the Cauchy problem, we mention here that Tonelli's approach was also applied by Cazzani Nieri [10] to the mixed problem for systems of the type (15). Indeed, assume that $1 \leqq k < m$ and that the following inequalities hold

$$\lambda_i(x,y;\ z_1,\dots,z_m) > 0, \qquad (i = 1,\dots,k),$$

$$\lambda_i(x,y;\ z_1,\dots,z_m) < 0, \qquad (i = k+1,\dots,m),$$

$$|\lambda_i(x,y;\ z_1,\dots,z_m)| \leqq L(x), \quad (i = 1,\dots,m).$$

Then, given m+k functions

$$\begin{aligned} &\phi_i(y),\ (i = 1,\dots,m),\ \text{Lipschitzian in } (0,b_o),\\ &\Psi_j(x),\ (j = 1,\dots,k),\ \text{absolutely continuous in } (0,a_o), \end{aligned} \tag{22}$$

with $\phi_j(0) = \Psi_j(0)$, $(j = 1,\dots,k)$, Cazzani Nieri proves the existence of a number a with $0 < a \leqq a_o$, such that in the region

$$T : 0 \leqq x \leqq a, \quad 0 \leqq y \leqq b_o - \int_0^x L(t)dt,$$

there exists at least one m-mple of functions $z_i(x,y)$, $(i = 1,\dots,m)$ of class G satisfying system (15) almost everywhere in T, and the requirements

$$z_i(0,y) = \phi_i(y), \quad (0 \leqq y \leqq b_o), \qquad (i = 1,\dots,m),$$

$$z_i(x,0) = \Psi_j(x), \quad (0 \leqq x \leqq a), \qquad (j = 1,\dots,k).$$

Though we refer to the paper [10] for details, we mention here how the "back step" method was applied. Let a_2 be an appropriate constant $0 < a_2 \leqq a_o$. Then, for every integer $i = 1,\dots,m$, Cazzani Nieri defines two sequences of functions $g_i^{[n]}(X;x,y)$, $z_i^{[n]}(x,y)$, $n = 2,3,\ldots$, by taking first $\delta_n = a_2/n$ and

$$z_i^{[n]}(x,y) = \phi_i(y), \quad \text{for } x \leq 0.$$

Now let $g_i^{[n]} = g_i^{[n]}(X;x,y)$ be the unique solution of the ordinary differential equation

$$g_i^{[n]} = y - \int_X^X \sigma_i^{[n]}(t,g_i^{[n]})dt$$

in the region

$$0 \leq X \leq a_2, \quad 0 \leq x \leq a_2, \quad -\infty < y < +\infty,$$

and where

$$\sigma_i^{[n]}(X,Y) = \lambda_i(X,Y,z_1^{[n]}(X - \delta_n,Y),\ldots,z_m^{[n]}(\ldots))$$

for (X,Y) in the region

$$D_2 : 0 \leq X \leq a_2, \quad 0 \leq Y \leq b_0 - \int_0^X L(t)dt,$$

and

$$\sigma_i^{[n]}(X,Y) = \sigma_i^{[n]}(X,0) \quad \text{for } Y < 0;$$

$$\sigma_i^{[n]}(X,Y) = \sigma_i^{[n]}(X, b_0 - \int_0^Y L(t)dt)), \text{ for } Y > b_0 - \int_0^X L(t)dt.$$

If we now consider the curve

$$y = g_i^{[n]}(x;0,0), \tag{23}$$

we denote by $a_i^{[n]} \leq a_2$ the maximum value of x, for which the arc of (23) corresponding to $0 \leq x \leq a_i^{[n]}$ is completely contained in D_2, we denote by $D_i^{[n]}$ the region of all points of D_2 satisfying the relations

$$0 \leq x \leq a_i^{[n]}, \quad g_i^{[n]}(x;0,0) \leq y \leq b_0 - \int_0^X L(t)dt,$$

and by $f_i^{[n]}(\ldots)$ the expression

$$f_i^{[n]}(\ldots) = f_i(X;g_i^{[n]}(X;x,y),\ z_1^{[n]}(X-\delta_n,\ g_i^{[n]}(X;x,y)),\ldots,z_m^{[n]}(\ldots)),$$

then Cazzani Nieri shows that we have

$$z_i^{[n]}(x,y) = \phi_i(g_i^{[n]}(0;x,y)) + \int_0^x f_i^{[n]}(\ldots)dX,$$

in all of D_2 for $i = k+1,\ldots,m$, as well as in $D_i^{[n]}$ for $i = 1,\ldots,k$, and that we have

$$z_i^{[n]}(x,t) = \Psi_i(\xi_i^{[n]}) + \int_{\xi_i^{[n]}}^x f_i^{[n]}(\ldots)dX$$

in all of $D_2-D_i^{[n]}$, where $\xi_i^{[n]}$ is the unique point with $0 < \xi_i^{[n]} < a_2$ where $g_i^{[n]}(\xi_i^{[n]};x,y) = 0$.

(b) Cazzani Nieri ([11], 1970; [14], 1980) studied further the case of two independent variables, and still under assumption (22). She proved the existence of at least one solution of class G for systems (17*) and (14*). In the proof, direct use is made of the approximations mentioned here.

13. (a) Let us return to the Cauchy problem of no. 10. Also in view of no. 11 (b), it is quite natural to ask the question whether it is possible to enlarge either the assumptions of the functions ϕ_i which determine the initial data, or the class G in which there is at least one solution to the problem.

(b) In the case in which there are only two independent variables, Cazzani Nieri [12] has given an answer to these questions for the semilinear system (17*) and the initial data (18*).

Under the assumption that the functions $\phi_j(y)$ are absolutely continuous, Cazzani Nieri has proved the existence of at least one solution of the class $H^{[0]}$ of all functions $z_j(x,y)$, $(j = 1,\ldots,m)$, which are absolutely continuous of x alone in a uniform way with respect to y, and absolutely continuous of y alone in a uniform way with respect to x. The assumptions on the functions a_{ij}, λ_i, f_i are particularizations of those of no. 10(b), but for the following two which deserve some attention. The question is whether the interval ($0 \leq x \leq a$) can be divided into a finite number of parts in each of which, for almost all x and for all y, one of the relations holds

$$\lambda_i(x,y) \geq 0, \quad \text{or } \lambda_i(x,y) \leq 0, \quad (i = 1,\ldots,m), \tag{24}$$

as well as one of the relations holds

$$\lambda_i(x,y) \leq \lambda_h(x,y), \text{ or } \lambda_i(x,y) \geq \lambda_h(x,y) \tag{25}$$

for every pair i, h, $i \neq h$, $(i,h = 1,\ldots,m)$.

For the proof, Cazzani Nieri introduced an auxiliary system J of integral equations, showing that the solutions of class $H^{[0]}$ of system J are also solutions of problem (17*) - (18*), and that such solutions are absolutely continuous as functions of only one variable along the characteristic lines of system (17*).

(c) Before we discuss conditions (24), (25), we wish to mention another paper by Cazzani Nieri [13] in which, in 1979, she proved a uniqueness theorem for the solution of problem (17*) - (18*). Indeed she proved uniqueness by preserving the requirement that the functions $\phi_j(y)$ are absolutely continuous, and by seeking the solutions in a class slightly more general than $H^{[0]}$, namely the class H of all functions $z_j(x,y)$, $(j = 1,\ldots,m)$, which are absolutely continuous in x for every y, and absolutely continuous in y uniformly with respect to x. The proof of this uniqueness theorem is completely independent from conditions (24), (25), and makes use of that process of approximation by integral means which Cinquini had already used for system (14) (cf. no. 11(b)).

It is important to add the remark here, that Cazzani Nieri, by means of the same approximation scheme, has shown in paper [13], that for solutions of class H, the integral system J and system (17*) - (18*) are equivalent.

(d) Let us now return to conditions (24), (25). Already in [12] Cazzani Nieri had shown, by means of two examples, that, if either (24) or (25) is not satisfied, then a possible solution of J may either not belong to the class $H^{[0]}$, or belongs to the class $H^{[0]}$ without being absolutely continuous along the characteristic lines of system (17*).

On the other hand, by the result in [13] mentioned at the end of (c) above, Cazzani Nieri showed that, if (24) is not satisfied, then there can exist no solution of class $H^{[0]}$ of problem (17*) - (18*).

The analogous question for condition (25) is still open since, for the time being, it was not possible to show that under all the remaining assumptions, problem (17*) -(18*) has a solution of class $H^{[0]}$, even if (25) is not satisfied; nor was it possible to construct a counterexample.

14. Conditions (24), (25) give rise to another question, namely, the question as to whether conditions (24), (25) can be disregarded if we assume that the functions $\phi_j(y)$, which define the initial data, are absolutely continuous with $|\phi_j'(y)|^{\nu}$ integrable for some $\nu > 1$.

This question is suggested by the possibility of making use, for instance, of the theorems proved by Tonelli in 1933, in connection with double integrals of the calculus of variations, to prove the equicontinuity of a sequence of functions of two variables. As it concerns condition (24), the answer is negative, as shown by the example in no. 15. This example has some analogy with that of Cazzani Nieri [12].

On the other hand, as it concerns condition (25), the question is still open.

15. (a) For $1 < \mu < 2$ let

$$g(x) = (a-x)^{\mu} \sin \frac{a\pi}{a-x}, \quad 0 \leqq x < a, \quad g(a) = 0.$$

Then $g(x)$ is absolutely continuous for $0 \leqq x \leqq a$, and $g(0) = 0$. Let us consider now the Cauchy problem (26) - (27):

$$\frac{\partial z}{\partial x} + \lambda(x) \frac{\partial z}{\partial y} = f(x), \tag{26}$$

with $\lambda(x) = g'(x)$, and

$$z(0,y) = \phi(y) \equiv |y|^{\alpha} \tag{27}$$

with

$$0 < \alpha < 1, \quad \alpha\mu \leqq 1 \tag{28}$$

while $f(x)$, $(0 \leqq x \leqq a)$ is quasi-continuous and integrable.

Obviously, $|\phi'(y)|^{\nu}$ is integrable for $(1-\alpha)\nu < 1$, and therefore we can assume ν large, provided we take α sufficiently close to one. It is elementary to see that for any M with

$$M \geqq (1 + \frac{\pi}{\mu-1})a^{\mu},$$

then the function

$$z(x,y) = |y-g(x)|^{\alpha} + \int_0^x f(t)dt \tag{29}$$

is a solution of problem (26) - (27) in the region $0 \leqq x \leqq a$, $-M \leqq y \leqq M$.

(b) If we take, for $0 \leqq x \leqq a$,

$$L(x) = \mu(a-x)^{\mu-1} + \frac{a\pi}{(a-x)^{2-\mu}},$$

then $|g'(x)| \leqq L(x)$. Thus, if $a_1 < a_2 < \ldots < a_k < \ldots$ is the sequence of all points in $0 < x < a$ where $g'(x) = 0$, then, since $a_k \to a$ as $k \to \infty$, we see that, for any fixed number $0 < a' < a$, and by the existence and uniqueness theorems of Cazzani Nieri (mentioned in no. 13), the function (29) is the unique solution of class $H^{[0]}$ of problem (26) - (27) in the region $T' : 0 \leqq x \leqq a'$, $-M + \int_0^x L(t)dt \leqq y \leqq M - \int_0^x L(t)dt$.

(c) However, in the region

$$T : 0 \leqq x \leqq a, \quad -M + \int_0^x L(t)dt \leqq y \leqq M - \int_0^x L(t)dt,$$

the function (29) is not of class $H^{[0]}$, since for $y = 0$ such a function is not even of bounded variation in $0 \leqq x \leqq a$, as we can verify immediately by taking note of the second relation (28).

Furthermore, in the region T there cannot be any solution of class $H^{[0]}$ of the Cauchy problem under consideration, since, as we have seen in (b), any possible solution of class $H^{[0]}$ should agree with (29) in the entire region T' with $a' < a$ and as close to a as we want.

(d) It is clear that, if instead of the first relation (28) we take $\alpha = 1$, hence $\phi(y)$ is Lipschitzian, then in the whole region T problem (26) - (27) has the solution (29) which, for $\alpha = 1$, belongs to the class G.

(e) It is clear that the assumption though restrictive that the initial data are given by Lipschitzian functions ϕ_j leads easily to strong conclusions. On the other hand, from the viewpoint of Cazzani Nieri, further work is conceivable: for instance, in the semilinear case, one could explore the question of the initial data when the independent variables are r+1 ($r > 1$). The quasi-linear case may conceivably present further difficulties.

On July 11, 1946, that is, only four months after the demise of Tonelli, I stated [16] that "though his major and original work has been in the calculus of variations, the complex of his further contributions in other fields of mathematics, particularly those in the theory of functions of real variables, would already assure him a place among the great". I hope that those gathered here at the Accademia delle Scienze dell'Istituto di Bologna will confirm my statement of 39 years ago: after the demise of Tonelli, one of his constructive processes has led to a great deal of research, and to a body of remarkable results in the theory of partial differential equations.

The work discussed in this paper was done in the frame of G.N.A.F.A.

REFERENCES

[1] G. Arnese, *Sull'approssimazione, col metodo di Tonelli, delle soluzioni del problema di Darboux per l'equazione* $s = f(x,y,z,p,q)$. Ricerche di Matematica, 11 (1962), 61-75.

[2] G. Arnese, *Sul problema di Darboux in ipotesi di Carathéodory*. ibidem, 12 (1963), 13-31.

[3] E. Baiada, *Sul teorema d'esistenza per le equazioni alle derivate parziali del primo ordine*. Ann. Scuola. Norm. Sup. Pisa, 12 (1943) [1947], 135-145.

[4] E. Baiada, *Considerazioni sull'esistenza della soluzione per un'equazione alle derivate parziali, con i dati iniziali, nel campo reale*. Ann. Mat. p. e appl., 34 (1953), 1-25.

[5] E. Baiada, C. Vinti, *Un teorema d'esistenza della soluzione per un' equazione alle derivate parziali del 1° ordine*. Ann. Scuola. Norm. Sup. Pisa, 9 (1955), 115-140.

[6] P. Bassanini, *Iterative methods for quasi linear hyperbolic systems*. Boll. Un. Mat. It., 1 B (1982), 225-250.

[6i] P. Bassanini, L. Cesari, *La duplicazione di frequenza nella radiazione laser*. Rend. Accad. Naz. Lincei 69 (1980), 166-173.

[6ii] P. Brandi, *Sul problema di Cauchy per una equazione differenziale nonlineare alle derivate parziali del primo ordine con argomento funzionale.* Atti Sem. Mat. Fis. Univ. Modena 28 (1979), 86-101.

[6iii] P. Brandi, R. Ceppitelli, *On the existence of the solution of non-linear functional partial differential equations of the first order.* ibidem 29 (1980), 166-186.

[7] D. Castaldo, *Su una generalizzazione di un problema di Darboux.* Rend. Circ. Mat. Palermo, 17 (1968), 240-248.

[8] L. Castellano, *Sull'approssimazione, col metodo di Tonelli, delle soluzioni del problema di Darboux per l'equazione* $u_{xyz} = f(x,y,z,u,u_x,u_y,u_z)$. Le Matematiche, 23 (1968), 107-123.

[9] M.G. Cazzani Nieri, *Un teorema di esistenza per un sistema di equazioni a derivate parziali del primo ordine.* Rend. Ist. Lomb. Accad. Sci. e Lett., 97 (1963), 455-481.

[10] M.G. Cazzani Nieri, *Su un problema misto per un sistema di equazioni a derivate parziali.* Ann. Mat. p. e appl., 77 (1967), 131-178.

[11] M.G. Cazzani Nieri, *Un teorema di esistenza per un problema misto.* ibidem, 87 (1970), 163-226.

[12] M.G. Cazzani Nieri, *Un teorema di esistenza della soluzione del problema di Cauchy nella classe delle funzioni assolutamente continue nelle singole variabili.* ibidem, 113 (1977), 127-171.

[13] M.G. Cazzani Nieri, *Teorema di unicitá e proprietá della soluzione del problema di Cauchy nella classe delle funzioni assolutamente continue nelle singole variabili.* Ricerche di Matematica, 28 (1979), 215-258.

[14] M.G. Cazzani Nieri, *Un teorema di esistenza per sistemi di equazioni a derivate parziali quasi lineari.* ibidem, 29 (1980), 17-64.

[14i] L. Cesari, *A boundary value problem for quasilinear hyperbolic systems in the bicharacteristic canonic form.* Ann. Scuola Norm. Sup. Pisa 1 (1974), 311-358.

[15] S. Cinquini, *Sulle equazioni funzionali del tipo di Volterra.* Rend. R. Accad. Naz. Lincei, 17 (1933), 616-621.

[16] S. Cinquini, *Leonida Tonelli.* Rend. Ist. Lomb. Sci. e Lett., 79 (1945-46), Parte gen. 73-93.

[17] S. Cinquini, *Sopra l'unicitá della soluzione per sistemi di equazioni a derivate parziali del primo ordine.* Rend. Ist. Lomb. Accad. Sci. e Lett., 88 (1955), 960-978.

[18] S. Cinquini, *Un teorema di unicitá per sistemi di equazioni a derivate parziali quasi-lineari.* Ann. Mat. p. e appl., 75 (1967), 231-260.

[19] S. Cinquini, *Sopra alcuni aspetti della riduzione a forma caratteristica dei sistemi di equazioni a derivate parziali.* Atti Conv. celeb. 80° ann. nascita R. Calapso, Messina-Taormina, 1981, 92-112.

[20] S. Cinquini, *Sopra la forma caratteristica dei sistemi di equazioni a derivate parziali di tipo iperbolico.* Rend. Ist. Lomb. Accad. Sci. e Lett., 117 (1983), 213-221.

[21] M. Cinquini Cibrario, *Nuovi teoremi di esistenza e di unicitá per sistemi di equazioni a derivate parziali.* Ann. Scuola. Norm. Sup. Pisa, 9 (1955), 65-113.

[22] M. Cinquini Cibrario, *Sistemi di equazioni a derivate parziali in piú variabili indipendenti.* Ann. Mat. p. e appl., 44 (1957), 357-418.

[23] M. Cinquini Cibrario, *Teoremi di esistenza per sistemi semilineari di equazioni a derivate parziali in piú variabili indipendenti.* ibidem, 68 (1965), 119-160.

[24] M. Cinquini Cibrario, *Teoremi di esistenza per sistemi di equazioni quasi lineari a derivate parziali in piú variabili indipendenti.* ibidem, 75 (1967), 1-46.

[25] M. Cinquini Cibrario, *Alcune recenti ricerche relative ai sistemi di equazioni a derivate parziali.* Atti Conv. celeb. 80° ann. nascita R. Calapso, Messina-Taormina, 1981, 76-91.

[26] M. Cinquini Cibrario, S. Cinquini, *Equazioni a derivate parziali di tipo iperbolico.* Ed. Cremonese, 1964, pp. VIII+552.

[27] R. Conti, *Sul problema iniziale per i sistemi di equazioni alle derivate parziali* $z_x^{(i)} = f^{(i)}(x,y;z^{(1)},\dots,z^{(k)};z_y^{(i)})$. Note I e II. Rend. Accad. Naz. Lincei, 12 (1952), 61-65 e 151-155.

[28] R. Conti, *Sul problema di Darboux per l'equazione* $z_{xy} = f(x,y,z,z_x,z_y)$. Ann. Univ. Ferrara, 2 (1953), 129-140.

[29] R. Conti, *Sull'equazione integrodifferenziale di Darboux-Picard.* Le Matematiche, 13 (1958), 30-39.

[30] N. Fedele, *Sull'approssimazione, col metodo di Tonelli, delle soluzioni di certi problemi relativi ad equazioni non lineari di tipo iperbolico.* Rend. Accad. Sci. Fis. Mat. Napoli, 31 (1964), 7-29.

[31] N. Fedele, *Sull'integrale superiore e quello inferiore di alcuni problemi relativi ad equazioni non lineari di tipo iperbolico, in ipotesi di Carathéodory.* Le Matematiche, 21 (1966), 167-197.

[32] D. Graffi, *Sopra una equazione funzionale e la sua applicazione a un problema di fisica ereditaria.* Ann. Mat. p. e appl., 9 (1931), 143-179.

[32i] D. Graffi, *Nonlinear partial differential equations in physical problems.* Pitman Adv. Publ. Program, 1980.

[33] F. Guglielmino, *Sulla risoluzione del problema di Darboux per l'equazione* $s = f(x,y,z)$. Boll. Un. Mat. It., 13 (1958), 308-318.

[34] F. Guglielmino, *Sull'esistenza delle soluzioni dei problemi relativi alle equazioni non lineari di tipo iperbolico in due variabili.* Le Matematiche, 14 (1959), 67-80.

[35] E. Magenes, *Sopra un problema di T. Satô per l'equazione differenziale* $y'' = f(x,y,y')$. Note I e II. Rend. Accad. Naz. Lincei, 2 (1947), 130-136 e 258-261.

[36] L. Merli, *Un problema ai limiti per una classe di sistemi di equazioni integrali.* Ann. Mat. p. e appl., 51 (1960), 139-146.

[37] R. Nardini, *Studio e risoluzione di un'equazione funzionale del tipo misto.* Ann. R. Scuola Norm. Sup. Pisa, 9 (1940), 201-213.

[38] M. Pagni, *Un'osservazione sull'unicità della soluzione del problema di Cauchy per l'equazione* $p = f(x,y,z,q)$. Rend. Sem. Mat. Univ. Padova, 20 (1951), 470-474.

[39] G. Pulvirenti, *Il fenomeno di Peano nel problema di Darboux per l'equazione* $s = f(x,y,z)$ *in ipotesi di Carathéodory.* Le Matematiche 15 (1960), 15-28.

[39i] A. Salvadori, *Sul problema di Cauchy per una struttura ereditaria di tipo iperbolico. Esistenza, unicità e dipendenza continua.* Atti Sem. Mat. Fis. Univ. Modena 32 (1983), 329-356.

[40] G. Sansone, *Equazioni differenziali nel campo reale.* Parte I e II; N. Zanichelli, 1941. (Vedi: I, 42; II, 130).

[41] G. Santagati, *Su alcuni sistemi di equazioni integro-differenziali in ipotesi di Carathéodory.* Ann. Mat. p. e appl., 67 (1965), 235-300.

[42] G. Santagati, *Sul problema di Picard in ipotesi di Carathéodory.* ibidem, 75 (1967), 47-94.

[43] G. Stampacchia, *Sulle condizioni che determinano gli integrali di un sistema di due equazioni differenziali ordinarie del primo ordine.* Rend. Accad. Naz. Lincei, 2 (1947), 411-418.

[44] G. Stampacchia, *Sulle condizioni che determinano gli integrali dei sistemi di equazioni differenziali del primo ordine.* Giorn. Mat. Battaglini, 77 (1947), 55-60.

[45] L. Tonelli, *Sulle equazioni integrali di Volterra.* Mem. R. Accad. Sci. Bologna, 5 (1927-28), 59-64. [Opere scelte, 4, 170-177.]

[46] L. Tonelli, *Sulle equazioni funzionali di Volterra.* Bull. Calcutta Math. Soc., 20 (1928), 31-48. [Opere scelte, 4, 198-212.]

[47] G. Villari, *Su un problema al contorno per una classe di sistemi di equazioni alle derivate parziali.* Boll. Un. Mat. It., 13 (1958), 514-521.

[48] C. Vinti, *Un teorema di esistenza per i sistemi di equazioni alle derivate parziali della forma* $p^{(i)} = F^{(i)}(x,y,z^{(1)},z^{(2)},\ldots,z^{(n)},q^{(i)})$. Atti Sem. Mat. Fis. Univ. Modena, 12 (1963), 33-106.

[49] W. Walter, *Über die Differentialgleichungen* $u_{xy} = f(x,y,u,u_x,u_y)$. Note I, II e III. Math. Zeitschrift, 71 (1959), 309-324 e 436-453; 73 (1960), 268-279.

[50] W. Walter, *On the existence theorem of Carathéodory for ordinary and hyperbolic equations.* Technical note BN-172, AFOSR (1959).

[51] A. Zitarosa, *Alcune osservazioni su certi sistemi di compattezza e sul problema di Darboux.* Rend. Accad. Sci. Fis. Mat. Napoli, 27 (1960), 25-35.

S. Cinquini
Dipartimento di Matematica
Università di Pavia
27100 Pavia
Italy.

F H CLARKE

Regularity, existence and necessary conditions for the basic problem in the calculus of variations

Leonida Tonelli's contributions to the theory of existence in the calculus of variations represent the best-known part of his work. In this article we seek to put into perspective certain other parts, those bearing upon regularity and upon necessary conditions, and demonstrate how together the development of these themes constitute a watershed between the classical (inductive) approach of the "naive theory" (in the words of L.C. Young) and the deductive (direct) methods in the calculus of variations.

In addition to reviewing certain aspects of the classical theory in order to be able to place these lesser-known contributions of Tonelli in context, we shall also describe some quite recent results on these issues, some of which have been directly inspired by Tonelli's work.

1. THE NAIVE THEORY

Consider the basic problem (P) in the calculus of variations: that of minimizing

$$J(x) := \int_{[a,b]} L(t,x(t),x'(t))dt$$

over a class X of functions $x(\cdot) : [a,b] \to R^n$, and subject to boundary conditions $x(a) = A$, $x(b) = B$. One of the themes of this article concerns the evolution of the theory of this problem during the three centuries in which it has been intensively studied.

In the early period, it was often assumed that a solution to (P) exists (hence "naive"), and the emphasis was placed upon necessary conditions that would enable one to identify the solution. The basic one is that of Euler (1744): if L is C^2, if X is the class of twice continuously differentiable functions, then the solution x to (P) satisfies *Euler's equation* on [a,b]:

$$\text{(E)} \quad d/dt\ L_v(t,x,x') = L_x(t,x,x').$$

It is of interest to recall what is now the standard derivation of (E), using a technique (due to Lagrange) from which the subject derives its name. Let $h(\cdot)$ be C^2 and vanish at a and b (h is called a *variation*). Then define, for any real number λ, $g(\lambda) := J(x + \lambda h)$. Since $x + \lambda h$ is feasible for (P) for any λ, the function g attains a minimum at 0, whence $g'(0) = 0$. By switching derivative and integral we arrive at

$$g'(0) = 0 = \int_{[a,b]} \{L_x h + L_v h'\}dt, \tag{1.1}$$

where the partial derivatives L_x and L_v are evaluated at (t,x,x'). Integrating by parts on the second term yields

$$0 = \int_{[a,b]} \{L_x - (d/dt)L_v\}h(t)dt. \tag{1.2}$$

Since this is true for any variation $h(\cdot)$, it follows that the term in braces vanishes, which is (E).

It is not too soon to already identify a link between two of the main themes under discussion: necessary conditions and regularity. To begin, note that an a priori regularity (smoothness) assumption on the solution x plays an important role in the derivation of the Euler equation (that is, differentiation under the integral, integration by parts, etc. require certain technical hypotheses). On the other hand, there is a tradition of proving regularity results via necessary conditions. To illustrate, we cite this classic proposition of Hilbert: if L is C^2 $(r \geqq 2)$ and $L_{vv} > 0$ (along x), then x belongs to C^r. The proof (see Cesari [2]) proceeds by an analysis of the Euler equation.

The first appearance of nonsmoothness in the calculus of variations may have been the recognition that the unique solution to (P) may admit corners (i.e., points of nondifferentiability), even when L is a polynomial; we refer to [2] for examples. For this reason, one often finds the classical theory developed with X = PWS, the class of piecewise-smooth functions. The choice of class of functions necessarily plays an important role in the necessary conditions. For example, consider the Euler equation (E) for the version of (P) in which $L = v^2$. The Euler equation is $x'' = 0$, which in C^2 leads to a unique possibility (in light of the boundary conditions). But in PWS, the condition $x'' = 0$ (which holds except at corner points) says only that x is one of infinitely many piecewise-linear admissible functions. The solution to this difficulty is to obtain additional necessary conditions. Erdmann (1877) put forward the following one:

> If x in PWS solves (P), then the function $t \to L_v(t,x(t),x'(t))$ is continuous.

Applied to the example above, this condition affirms that x' is continuous: i.e., that x has no corner points. Erdmann based his work on the work of Weierstrass, but later duBois-Reymond (1879) found a more direct approach: Go back to (1.1), and instead of using integration by parts on the second term, use it on the first. We obtain

$$0 = \int_{[a,b]} \{- \int_{[a,t]} L_x + L_v\}h'(t)dt.$$

It follows now from the arbitrariness of the variation h that the quantity in braces is constant. We refer to this as the integral form of the Euler equation:

$$\text{(1E)} \quad L_v(t,x,x') = c + \int_{[a,t]} L_x(\tau,x,x')d\tau.$$

Clearly (1E) implies both (E) and the Erdmann condition stated above.

We shall not go into further detail concerning classical developments, but we remark that like the results above, they generally extend without difficulty to certain larger classes of functions X, notably LPS, the class of absolutely continuous functions on [a,b] having essentially bounded derivative (equivalently, the class of x admitting a Lipschitz constant K on [a,b] : $|x(s) - x(t)| \leq K|s-t|$ for all s,t in [a,b].). Of course, the necessary machinery for dealing with such functions did not exist prior to Lebesgue. We shall see that the situation begins to change dramatically when the class X contains functions x having unbounded derivative.

2. DIRECT METHODS

To explain some of the significance of Tonelli's existence theory, let us pause to discuss the two principal approaches to solving optimization problems: the inductive and the deductive methods. The deductive method proceeds through the following logical steps: a solution to the problem exists (existence), any solution satisfies such-and-such conditions (necessary conditions), an analysis of these conditions, and perhaps elimination, identifies a sole candidate x; therefore x is the solution.

When this method applies, it is very satisfying; we have found the solution through sheer logic and impeccable circumstantial evidence. Note however the reliance of the logical chain upon the first step: existence. Even if for a "reasonable-looking" problem the necessary conditions identify a unique x, we cannot conclude that x is the solution unless we know a priori that a solution really does exist. The lack of existence theorems in the classical (naive) calculus of variations explains why its problems are solved by the other (less direct) principal approach, to which we now turn.

The inductive method begins with a specific candidate x which we suspect of being the solution (inductive: from the specific to the general). This x may have been arrived at by intuition, experience, or necessary conditions, but in any case its optimal nature is verified by the application of *sufficient* conditions. The latter constitute a mechanism by which the fact that a specific x solves the problem is confirmed; the analog would be to producing an eyewitness.

Tonelli (1915) provided the first general existence theorem for (P). In this seminal work, in which are anticipated such related concepts as coercivity, convexity, lower semicontinuity and weak convergence, he postulated the following hypotheses on L (besides twice continuous differentiability): for all (t,x,v) one has

(T1) $L_{vv}(t,x,v) \geqq 0$

(T2) $L(t,x,v) \geqq \varepsilon|v|^2 + \alpha$, where $\varepsilon > 0$.

But just as important as the assumptions on L was his choice for X:AC, the class of absolutely continuous functions on [a,b]. The result, *Tonelli's existence theorem*, asserts that (P) admits a solution when X = AC and L satisfies (T1), (T2).

Two basic questions now come to mind. The first stems from the introduction of AC: is it really necessary? After all, the theory becomes much more difficult when x lies in AC. For example, the derivation of the Euler equation presented in §1 runs into difficulties: the steps can no longer be justified. Is this a technical phenomenon or does it reflect a fundamental difficulty? And in any case, can the solution x really be ill-behaved (for example, have unbounded derivative), or will it always lie in LPS (say)?

The second question relates to the deductive method. Now that the first step can be performed rigorously, what is the situation regarding the next one; i.e., the necessary conditions? In particular, what necessary conditions hold under merely those hypotheses guaranteeing existence?

Not surprisingly, Tonelli was quite aware of the significance of these issues, and did manage to address them to some extent. He conjectured that it was possible under (T1) and (T2) for the solution to admit unbounded derivative, a conjecture that was to be verified only in 1984 (see §3).

On the issue of necessary conditions, Tonelli proved the following, under hypotheses that the inequality in (T1) holds *strictly*, and that the dimension $n = 1$: there is an open subset Ω of full measure in $[a,b]$ in which the solution x is C^2 and satisfies the Euler equation. We refer to this as *Tonelli's regularity theorem.*

Notice an apparent anomaly here: the necessary condition for $X = AC$ is *weaker* than in the case $X = PWS$ (for example, the integral form (1E) of the Euler equation is not affirmed), whereas in passing from C^2 to PWS (see §1) *additional* necessary conditions were derived. The explanation of this is that optimality relative to AC is quite distinct from optimality relative to PWS (say), in contrast to the passage from C^2 to PWS. Specifically it can be shown that (for the same boundary conditions) one has

$$\inf_{PWS} J = \inf_{C^2} J = \inf_{LPS} J$$

but, as Lavrentiev was the first to show (see [2]), it is possible to have

$$\inf_{AC} J < \inf_{LPS} J$$

even for L a polynomial. This is a surprising fact if one considers the many ways that a function x' in $L^1[a,b]$ can be approximated by a function in $L^\infty[a,b]$.

We remark that one way to salvage the deductive approach for (P) is to posit *additional* hypotheses on the Lagrangian L which will imply after the fact (i.e., after applying the existence theory on AC) that the solution x lies in LPS, and which therefore render accessible the gamut of necessary conditions. The condition one most often finds in this regard was suggested by Tonelli: a global growth hypothesis on the derivatives of L of the form

$$|L_x| + |L_v| \leqq c|L| + k. \qquad (2.1)$$

In addition to being in common use in the calculus of variations (see for example [10]), this condition is often seen in optimal control theory (see for example [3]).

3. GENERALIZED SOLUTIONS OF THE HAMILTON-JACOBI EQUATION

As we have said in §2, the inductive method proceeds through verification of the optimality of some predetermined candidate. We now proceed to describe one such technique in the calculus of variations - the Hamilton-Jacobi verification technique - and how, once again, regularity plays a central role.

Suppose that z is a feasible arc for (P), and that some continuously differentiable function $\phi(t,x)$ exists which satisfies globally the inequality

$$\text{(HJ)} \quad L(t,x,v) \geqq \phi_t(t,x) + \phi_x(t,x) \circ v,$$

with equality holding when (t,x,v) equals (t,z(t),z'(t)). It then follows easily that z is the global solution to the problem, for if y is any feasible arc, substituting (x,v) = (y(t),y'(t)) in (HJ) and integrating gives

$$J(y) \geqq \int_{[a,b]} (d/dt)\, \phi(t,y(t))dt$$

$$= \phi(b,B) - \phi(a,A),$$

which provides a lower bound for J, one which is attained when y = z.

When the technique just described applies, we say that ϕ *verifies* the optimality of z. A principal advantage of this approach is the verification of a global rather than local minimum; another is how easily it adapts to other problems, such as those incorporating additional constraints. Of course, we are led to wonder whether a verification function can always be expected to exist, and how to produce one.

The fact is that even for classical variational problems, there may be no verification function ϕ as described above which confirms the optimality of a given solution. However, the technique may be extended to allow nonsmooth ϕ through the generalized gradient $\partial\phi$ (see [3]). In these terms, (HJ) becomes

$$L(t,x,v) \geqq \alpha + \beta \circ v \text{ for all } (\alpha,\beta) \in \partial\phi(t,x).$$

It follows much as before (but using some generalized calculus) that provided an additional compatibility property is satisfied by $x(\cdot)$, then ϕ confirms the optimality of $x(\cdot)$. When extended this way, the technique becomes necessary and sufficient [3, Section 3.7]: x is optimal for (P) iff there is a (Lipschitz) verification function ϕ. Once again then, and this time in an inductive setting, regularity plays a central role; it is necessary to consider nonsmooth functions even if our interest is limited to smooth problems.

The nonsmooth version of the Hamilton-Jacobi technique just described can be used to establish the conjecture of Tonelli that the solution x to (P) can have unbounded derivative. It was employed by Clarke and Vinter [8] to confirm the putative solution to a specific problem first put forward by Ball and Mizel [1]. Note the need here of a technique yielding a global minimum.

We remark that the method described in this section can be cast in Hamiltonian terms by means of a generalized notion of solutions to the Hamilton-Jacobi equation. Specifically, let us recall the Hamiltonian H corresponding to L:

$$H(t,x,p) := \sup_{v \in R^n} \{p \circ v - L(t,x,v)\}.$$

Then the Hamilton-Jacobi inequality given above (for a nonsmooth function ϕ) can be replaced in the theory by:

$$\max_{(\alpha,\beta)\in\partial\phi(t,x)} \{\alpha + H(t,x,\beta)\} = 0.$$

When ϕ is smooth, this reduces to the classical Hamilton-Jacobi equation

$$\phi_t + H(t,x,\phi_x) = 0.$$

As shown in [3] for the more general setting of (normal) control problems a necessary and sufficient condition for optimality is the existence of a Lipschitz verification function satisfying the (generalized) Hamilton-Jacobi equation. As shown by Zeidan [11] (see also [12]), other types of sufficient conditions, such as convexity and those postulating the absence of conjugate points, can be viewed and obtained as special cases of this approach.

4. REGULARITY OF SOLUTIONS TO THE BASIC PROBLEM

The preceding sections underline the importance of regularity theorems to the deductive approach; without them the full range of necessary conditions are unavailable. In fact we now see that the crucial assertion is that the solution be Lipschitz; with this in hand, other regularity properties available in special cases follow easily.

Tonelli's regularity theorem cited in §2 inspired the following result of Clarke and Vinter [6]. The dimension n is now unrestricted, and L need not be smooth. The precise hypotheses are:

(H1) $L(t,x,v)$ is locally bounded on $[a,b] \times R^n \times R^n$, measurable as a function of t, and convex as a function of v.

(H2) L is locally Lipschitz in (x,v) uniformly in t; i.e., for each bounded subset C of $R^n \times R^n$, there exists a constant K such that, for all (x_1,v_1), (x_2,v_2) in C, for all t in [a,b], one has

$$|L(t,x_1,v_1) - L(t,x_2,v_2)| \leqq K|(x_1 - x_2, v_1 - v_2)|.$$

(H3) There is a constant α and a convex function $\theta:[0,\infty) \to R$ such that

$$L(t,x,v) \geqq - \alpha|x| + \theta(|v|)$$

for all (t,x,v) in $[a,b] \times R^n \times R^n$, where $\theta(r)/r \to \infty$ as $r \to \infty$.

Theorem 1 *Under* (H1) - (H3), *a solution* x *to the problem exists. Let* τ *in* [a,b] *be any point such that*

$$(\ast) \qquad \liminf_{\substack{s,t \to \tau,\, s \neq t \\ a \leqq s \leqq \tau \leqq t \leqq b}} |x(t) - x(s)|/(t - s) < \infty$$

Then there exists an interval I *which is a neighbourhood of* τ *in* [a,b] *in which the arc* x *is Lipschitz.*

Interestingly, and in keeping with our theme, necessary conditions and existence theory play a major role in the proof.

The existence of the set Ω of Tonelli's regularity theorem follows easily from this result; the Ball-Mizel problem confirms that in general Ω may not be taken to be all of [a,b]. Somewhat more about Ω can be said when L is polynomial-like; see Clarke and Vinter [7].

The main use of Theorem 1 has been in the identification of additional hypotheses on L that have the property of implying that the solution x to (P) lies in LPS. Three such classes of hypotheses are developed in [6]. One class generalizes the type of growth condition given by (2.1), and another bears upon the special case in which the Euler equation (E) can be written explicitly as a second-order differential equation. A third class includes the case in which (P) is autonomous (i.e., L has no explicit dependence on the t variable). We have therefore the following new result: in the autonomous case, under the hypotheses of the Tonelli existence theorem, the solution to the basic problem has essentially bounded derivative. This is surely a fact that would have pleased Tonelli.

5. A SUMMARY OF RESULTS FOR THE AUTONOMOUS BASIC PROBLEM

Consider now the case of (P) in which the Lagrangian L(x,v) is independent of t, locally Lipschitz in (x,v), convex as a function of v, and majorizes a function of the form $-\alpha|x| + \theta(|v|)$, where $\theta(r)/r \to \infty$ as $r \to \infty$. The basic theory of this problem, viewed as a problem of optimization, is now attaining a certain completeness after three centuries. Here are the principal results, in which existence, regularity, necessary conditions and sufficient conditions are all inextricably interwoven:

(1) A solution x relative to AC exists; x itself is Lipschitz

(2) There exists a Lipschitz function p satisfying [3]

  (a) $(p'(t), p(t)) \in \partial L(x(t),x'(t))$ a.e. (Euler inclusion)

  (b) $p(t)\cdot x'(t) - L(x(t),x'(t)) = \text{constant}$ (second Erdmann condition)

  (c) $L(x(t),x'(t) + w) - L(x(t),x'(t)) \geqq p(t)\cdot w \;\forall\; w$, a.e. (Weierstrass condition)

(3) If L is strictly convex in v then x is C^1; if L is differentiable in v then $t \to L_v(x(t),x'(t))$ is Lipschitz (first Erdmann condition)

(4) If L is C^2 and $L_{vv}(x(t),x'(t))$ is positive definite, then x is C^2 and admits no points in (a,b) conjugate to a (Jacobi condition)

(5) If L and $L_{vv}(x(t),x'(t))$ are as in (4) where x is a C^2 extremal admitting no points in (a,b] conjugate to a, then x is a strong local minimum [9]

(6) The existence of a Lipschitz solution of the extended Hamilton-Jacobi equation compatible with x is a necessary and sufficient condition for x to be optimal.

The connections between the central themes become still more elaborate when other issues are considered, such as their Hamiltonian formulations [3], or their relationship to convexity [11]. Other questions arise as well, such as existence "in piccolo" [5], first broached by Tonelli also. And much work remains to be done for descendants of the basic problem, such as those involving constraints [4] or control dynamics or multiple-integrals: perhaps centuries worth.

ACKNOWLEDGEMENTS

The support of the NSERC of Canada is gratefully acknowledged.

REFERENCES

[1] J. Ball and V. Mizel, One-dimensional variational problems whose minimizers do not satisfy the Euler-Lagrange equation, Arch. Rat. Mech. Anal., 90 (1985) 325-388.

[2] L. Cesari, *Optimization-Theory and Applications*, Springer-Verlag, New York (1983).

[3] F.H. Clarke, *Optimization and Nonsmooth Analysis*, Wiley Interscience, New York (1983).

[4] F.H. Clarke, Regularity of solutions to the isoperimetric problem, in Proceedings of the Int. Conf. on Calculus of Variations (Trieste 1985), Springer-Verlag, to appear.

[5] F.H. Clarke and R.B. Vinter, Existence and regularity in the small in the calculus of variations, J. Differential Equations 59 (1985) 336-354.

[6] F.H. Clarke and R.B. Vinter, Regularity properties of solutions to the basic problem in the calculus of variations, Trans. Amer. Math. Soc. 289 (1985) 73-98.

[7] F.H. Clarke and R.B. Vinter, Regularity of solutions to variational problems with polynomial Lagrangians, Bull. Polish Acad. Sci., in press.

[8] F.H. Clarke and R.B. Vinter, On the conditions under which the Euler equation or the maximum principle hold, Appl. Math. Optim. 12 (1984) 73-79.

[9] F.H. Clarke and V. Zeidan, Sufficiency and the Jacobi condition in the calculus of variations, Canad. J. Math., to appear.

[10] C.B. Morrey Jr., *Multiple Integrals in the Calculus of Variations*, Springer-Verlag, New York (1966).

[11] V.M. Zeidan, Sufficient conditions for the generalized problem of Bolza, Trans. Amer. Math. Soc. 275 (1983) 561-586.

[12] V.M. Zeidan, First and second order sufficient conditions for optimal control and the calculus of variations, Appl. Math. Optim. 11 (1984) 209-226.

F.H. Clarke
Centre de Recherches Mathématiques
Université de Montréal
Montréal
Quebec
Canada H3C 3J7.

I EKELAND

The calculus of variations: from minimization problems to critical point theory

There are two great books in the calculus of variations: Caratheodory's "Variationsrechrung" (1935) and Tonelli's "Fondamenti di Calcolo delle Variazioni" (1921-23). There is no doubt that the second one, though older than the first, has withstood the test of time much better. Unfortunately it has never been translated into English, which may explain why so much of its contents have been forgotten until recently.

This paper is an attempt to show some of Tonelli's ideas at work in today's calculus of variations, and also to show new ideas which have recently come into this field.

THE DIRECT METHOD IN THE CALCULUS OF VARIATIONS

Let us consider the classical Bolza problem in the calculus of variations:

$$(P)\qquad \begin{cases} \operatorname{Inf} \int_0^T L(t,x(t),\dot{x}(t))dt \\ x(0) = \xi_0,\ x(T) = \xi_1,\ \dot{x} \in L^2. \end{cases}$$

The basic existence result is the following:

Theorem Assume for $L:[0,T] \times \mathbb{R}^n \times \mathbb{R}^n \to \mathbb{R} \cup \{+\infty\}$ the following properties

(a) L is Borelian in (t,x,u), and lower semi-continuous in (x,u)

(b) $L(t,x,u) \geq a(t) + \phi(\|u\|)$, with $a \in L^1(0,T)$ and $\phi(s)s^{-1} \to +\infty$ when $s \to \infty$ in $\mathbb{R}$.

(c) L is convex in u for fixed (t,x).

Then (P) has a solution $\bar{x}$ (i.e. the infimum is attained at $\bar{x}$).

Proof Step 1 Let $x_n \in L^1$ be a minimizing sequence:

$$\int_0^T L(t,x_n,\dot{x}_n)dt \to \operatorname{Inf}(P).$$

Then, by estimate (b),

$$\int_0^T \phi(\|\dot{x}_n\|)dt \leq \text{constant}.$$

It follows from the compactness criterion of Dunford and Pettis that the sequence $\dot{x}_n$ is weakly compact in L^1.

$$\dot{x}_n \to \dot{x} \text{ in the } \sigma(L^1,L^\infty) \text{ topology}$$

$$x_n \to x \text{ uniformly on } [0,T].$$

Step 2 We now use Mazur's lemma (a corollary of the Hahn-Banach theorem; see [10]): there is a sequence of convex combinations z_k:

$$\dot{z}_k = \sum_{n \leq k} \alpha_{n,k} \dot{x}_n \text{ with } \alpha_{n,k} \geq 0 \text{ and } \sum_{n=1}^{k} \alpha_{n,k} = 1$$

such that:

$$\dot{z}_k \to \dot{x} \text{ in } L^1(0,T) \text{ and a.e.}$$

$$z_k \to x \text{ uniformly on } [0,T].$$

Taking advantage of assumption (c), one is then able to show:

$$\text{Inf (P)} = \lim_k \int_0^T L(t,x_k,\dot{z}_k)dt.$$

Using Fatou's lemma, we then get:

$$\lim_k \int_0^T L(b,z_k,\dot{z}_k)dt \geq \int_0^T L(t,x,\dot{x})dt$$

Hence:

$$\text{Inf (P)} = \int_0^T L(t,x,\dot{x})dt.$$

Since x obviously satisfies the boundary conditions, it is the desired minimizer. □

We refer to [4] or [10] for details. The basic ideas in the proof, the use of convexity and lower semi-continuity of integrals, are due to Tonelli, and this theorem was essentially known to him. See [10] for relaxation theory, when the integrand is no longer convex in u.

WHAT DID TONELLI KNOW THAT WE HAVE FORGOTTEN?

The next basic question, once the existence problem has been solved, is regularity. It is known, for instance, that if the minimizer $\bar{x}$ is Lipschitz, that is $\|\frac{d\bar{x}}{dt}\|_\infty < \infty$, then $\bar{x}$ satisfies the Euler-Lagrange equation

$$\frac{d}{dt}\frac{\partial L}{\partial u}(t,x,\dot{x}) = \frac{\partial L}{\partial x}(t,x,\dot{x})$$

or some ad hoc version of it if L is not differentiable. But when is $\bar{x}$ Lipschitz? And what happens if it is not? Both questions were raised by Tonelli [12], and his answers apparently forgotten until Cesari revived them and Ball showed their importance in continuum mechanics.

The basic fact to account for is the following. Consider the two optimization problems:

$$(P_1)\quad \begin{cases} \text{Inf} \int_0^T L(t,x,\dot{x})dt \\ x(0) = \xi_0,\ x(T) = \xi_1 \\ \dot{x} \in L^1 \end{cases}$$

and:

$$(P_2)\quad \begin{cases} \text{Inf} \int_0^T L(t,x,\dot{x})dt \\ x(0) = \xi_1,\ x(T) = \xi_1 \\ \dot{x} \in C^0 \end{cases}$$

In other words, x is absolutely continuous in the first problem, and continuously differentiable in the second one. Then, for an appropriate choice of L (satisfying of course conditions (a) (b) and (c)), we will have

$$\text{Inf}\ (P_1) < \text{Inf}\ (P_2).$$

This is known as the *Lavrentiev phenomenon*. As a consequence, the minimizer in problem (P_1) will not satisfy the Euler-Lagrange equation (even though L is smooth). Tonelli [12] gave partial regularity results, and Ball-Mizel [2], [3] have come up with a specific example, which has been very helpful in furthering the analysis. For instance, the Lavrentiev phenomenon cannot happen when L is time-independent: L(x,u). In that case, any minimizer has to be Lipschitz, and satisfies Euler-Lagrange.

WHAT DO WE KNOW THAT TONELLI DID NOT KNOW?

The most striking development in recent years has been the treatment of non-minimization problems in the calculus of variations. Such problems are very common in mathematical physics. For instance, the standard equation for a harmonic oscillator:

$$\begin{cases} \ddot{x} + x = 0 \\ x(0) = x(2\pi) \end{cases}$$

can be framed as a problem in the calculus of variations:

$$\begin{cases} \Phi'(x) = 0 & \Phi(x) = \int_0^{2\pi} [\frac{\dot{x}^2}{2} - \frac{x^2}{2}]dt \\ x \in E & E = H^1(\mathbb{R}/2\,\mathbb{Z}; \mathbb{R}^{2n}) \end{cases}$$

There is obviously a solution to the problem, but it cannot be gotten by minimizing or maximizing the function, . Other methods (here, linear analysis) must be used.

More generally, many nonlinear problems in classical mechanics or physics can be cast as boundary-value problems for Hamiltonian systems, typically:

$$\begin{cases} \dot{x} = JH'(x) \\ x(0) = x(T) \end{cases} \tag{1}$$

where $T > 0$ is prescribed, $H : \mathbb{R}^{2n} \to \mathbb{R}$ is a C^2 map, $H'(x)$ its gradient, and $J \in \mathcal{L}(\mathbb{R}^{2n})$ is the matrix

$$J = \begin{pmatrix} 0 & I \\ -I & 0 \end{pmatrix}$$

Theorem (least action principle). Problem (1) is equivalent to:

$$\begin{cases} \Phi'(x) = 0 & \Phi(x) = \int_0^T [\tfrac{1}{2}(J\dot{x},x) + H(x)]dt \\ x \in E & E = H^1(\mathbb{R}/T\mathbb{Z}; \mathbb{R}^{2n}) \end{cases} \tag{2}$$

For the proof, one simply writes down the Euler-Lagrange equations associated with $\Phi'(x) = 0$. It should be noted that the least action principle is a misnomer: the action integral, $\Phi(x)$, is unbounded from above and from below, and can be neither minimized nor maximized. The see this, simply substitute $x_k(t) = \exp(2k\pi Jt/T)$, and note that $\Phi(x_k) \to \pm\infty$ when $k \to \pm\infty$ in $\mathbb{Z}$.

The first breakthrough on problem (2) is due to Rabinowitz [11]. Since then, abstract results in critical point theory have found in this problem a host of applications. We will now describe a particular method, initiated by Clarke [5], applicable in the case when the Hamiltonian is convex.

CONVEXITY AND DUALITY

From now on, we assume that

$$\text{(A)} \qquad \begin{cases} H : \mathbb{R}^{2n} \to \mathbb{R} \text{ is convex} \\ H(x)\,\|x\|^{-1} \to +\infty \text{ when } \|x\| \to \infty \\ H(x) \geq H(0) = 0 \end{cases}$$

H then has a well-defined Legendre transform (or Fenchel conjugate) G, defined by:

$$G(y) = \text{Max}\,\{(x,y) - H(x) \mid x \in \mathbb{R}^{2n}\}$$

with the fundamental property (reciprocity formula)

$$y \in \partial H(x) \iff x \in \partial G(y).$$

If H and G are C^1, this last relation is simply $(H')^{-1} = G'$. Note that, even if H is not differentiable, we will be able to solve the boundary-value problem:

$$\begin{cases} \dot{x} \in J\partial H(x) \\ x(0) = x(T) \end{cases} \tag{3}$$

which replaces problem (1). For information on $\partial H(x)$, the subgradient of H, we refer to [3].

If $H(x) = \frac{1}{\alpha}\|x\|^{\alpha}$, then $G(x) = \frac{1}{\beta}\|x\|^{\beta}$, where $\alpha^{-1} + \beta^{-1} = 1$. Note that if

H is subquadratic ($1 < a < 2$), then β is superquadratic ($\beta > 2$), and conversely.

Theorem (dual action principle) Problem (3) is equivalent to:

$$\begin{cases} \psi'(x) = 0 & \psi(x) = \int_0^T [\tfrac{1}{2}(J\dot{x},x) + G(-J\dot{x})]dt \\ x \in E & E = H^1(\mathbb{R}/T\mathbb{Z};\mathbb{R}^{2n}) \end{cases} \tag{4}$$

This result is essentially due to Clarke [5]. The equivalence takes the following form: if x solves (3), then x solves (4), and if x solves (4), then some translate $x + \xi$ of x solves (3), with ξ a constant vector in $\mathbb{R}^{2n}$. The function ψ is the sum of a quadratic form and a convex l.s.c. function, and the equation $\psi'(x) = 0$ should be interpreted accordingly.

For the proof, simply write down the Euler-Lagrange equation, and use the reciprocity formula (see [6]).

As a functional, the dual action principle is much better behaved than the least action principle, because of the presence of the derivative $\dot{x}$ in the nonlinear term. As we shall see, it can even be minimized in some cases.

SUBQUADRATIC PROBLEMS

Theorem Assume that there exist two numbers k and K, with $0 < k < K$, such that

(a) $\liminf_{\|x\| \to 0} H(x) \|x\|^{-2} \geq \frac{K}{2}$

(b) $\limsup_{\|x\| \to \infty} H(x) \|x\|^{-2} \leq \frac{k}{2}$

Then, for all $T \in (\frac{2\pi}{K}, \frac{2\pi}{k})$, problem (3) has at least one solution with minimal period T.

This is a result of Clarke and Ekeland [6]. Any solution of problem (3) has to be T-periodic, and the theorem asserts that there is a solution which has no lower period than T; in particular, it cannot be the constant solution $x = 0$.

The proof goes by applying the direct method of the calculus of variations to the dual action principle:

$$\psi(x) = \int_0^T [\tfrac{1}{2}(J\dot{x},x) + G(-J\dot{x})]dt$$

on the space $H^1(\mathbb{R}/T\mathbb{Z};\mathbb{R}^{2n})$.

The true variable is $\dot{x}$ and not x (changing x to $x + \xi$ - does not change the value of ψ). The function ψ is the sum of two terms, the first one being compact (hence weakly continuous), the second one convex (hence weakly l.s.c.). Because of estimate (b), it is the second one which is dominant when $||\dot{x}|| \to \infty$, so ψ is coercive and attains its minimum. Because of estimate (a), it is the first one which is dominant near $\dot{x} = 0$; since it is indefinite, cannot attain its minimum at 0, so it must be attained elsewhere, and the solution we get for problem (3) cannot be the constant one.

SUPERQUADRATIC PROBLEMS

Theorem Assume that, for some $\alpha > 2$:

$$(H'(x),x) \geqq \alpha H(x).$$

Then for each $T > 0$, problem (3) has at least one solution with minimal period T.

The existence of a non-constant T-periodic solution was first proved by Rabinowitz [11]; the proof we shall give is Ekeland's [7]. The existence of a solution with minimal period T was first proved by Ekeland and Hofer [9].

Again one considers the dual action principle ψ. This time the role of the two terms is reversed. When $||\dot{x}|| \to \infty$, the first term is dominant, and it can have either sign. When $\dot{x}$ is near 0, the second term is dominant, and it is positive. So ψ has a local minimum at the origin, and points can be found near infinity where $\psi < \psi(0)$. It can be proved that ψ satisfies condition (c) of Palais and Smale, so that, by celebrated theorem of Ambrosetti and Rabinowitz ψ has a non-zero critical point (see [1]).

The minimal period question is more subtle and requires an incursion into index theory.

INDEX THEORY

Let us have another look at the dual action functional. Let x be a critical point, $\psi'(x) = 0$, and write down the Hessian of ψ at x:

$$(\psi''(x)y,y) = \int_0^T [Jy,y) + (G''(-J\dot{x})Jy,Jy)]dt.$$

This is a quadratic form on $H^1(\mathbb{R}/T\mathbb{Z};\mathbb{R}^{2n})$ which consists of two terms, the first one being compact and the second one positive definite. It follows that it must have finite index and finite nullity. In other words, there is a splitting

$$H^1(\mathbb{R}/T\mathbb{Z};\mathbb{R}^{2n}) = E_+ \oplus \ \text{Ker}\ \psi''(x) \oplus E_-$$

into $\psi''(x)$-orthogonal subspaces, positive definite, zero, and negative definite, the last two being finite-dimensional.

Definition The index of x is the dimension of E_-. The nullity of x is the dimension of Ker $\psi''(x)$.

We seek a geometrical interpretation of the index. In other words, until now we have seen x as a critical point of ψ. We now look at it as a T-periodic solution of $\dot{x} = JH'(x)$. The two points of view are equivalent, by the dual action principle, and should lead to the same results.

Definition We say that two points $x(s_1)$ and $x(s_2)$, with $s_1 < s_2$, are conjugate along the solution x, if the linear problem:

$$\begin{cases} \dot{y} = JH''(x(t))y \\ y(s_1) = y(s_2) \end{cases}$$

has a non-zero solution. The multiplicity is the number of linearly independent solutions.

We then have the geometrical interpretation we were seeking (see [8] or [9] for a proof).

Theorem The index of x is the number of $s \in (0,T)$ such that x(s) is conjugate to x(0), each counted with multiplicity.

MINIMAL PERIOD PROBLEMS

We now turn our attention to the boundary-value problem:

$$\begin{cases} \dot{x} = JH'(x) \\ x(0) = x(T) \end{cases}$$

Clearly, any solution x to this problem is T-periodic, and we want to see if it has some lower period.

Proposition If x has index i, its minimal period is T, T/2,..., or T/(i+1).

Proof It is well-known that $y = \dot{x}$ is a solution of the linearized system $\dot{y} = JH''(x(t))y$. If now x is T/k-periodic we will have $\dot{y}(0) = \dot{y}(T/k) = \ldots = \dot{y}(\frac{k-1}{k}T)$, so that $x(\frac{T}{k}),\ldots,x(\frac{k-1}{k}T)$ are all conjugate to x(0). This proves that the index must be at least k-1.

Corollary In the subquadratic case, the solution we found has minimal period T.

Proof We got x by minimizing ψ. So x must have index 0, and its minimal period can only be T.

In the superquadratic case, it can be shown that the index of the solution we get by the Ambrosetti - Rabinowitz theorem must be 0 or 1. So the minimal period can only be T or T/2. A direct argument eliminates T/2, so the minimal period is T. We refer to [9] for details.

REFERENCES

[1] J.P. Aubin and I. Ekeland, "Applied nonlinear analysis", Wiley 1984.

[2] J. Ball and V. Mizel "Singular minimizers for regular one-dimensional problems in the calculus of variations" Bull. AMS 11 (1984) p. 143-146.

[3] J. Ball and V. Mizel, "One-dimensional variational problems whose minimizers do not satisfy the Euler-Lagrange equation", preprint 1985.

[4] L. Cesari "Optimization: theory and applications", Springer 1983.

[5] F. Clarke, "Periodic solutions of Hamiltonian inclusions" J. Diff. Eq. 40 (1981) p. 1-6.

[6] F. Clarke and I. Ekeland, "Hamiltonian trajectories having prescribed minimal period" Comm. PAM 33 (1980) p. 193-116.

[7] I. Ekeland, "Periodic solutions to Hamiltonian equations and a theorem of P. Rabinowitz", J. Diff. Eq. 34 (1979) p. 523-534.

[8] I. Ekeland, "An index theory for periodic solutions of convex Hamiltonian systems", Proc. AMS Summer Institute on Nonlinear Functional Analysis, Berkeley 1983.

[9] I. Ekeland and H. Hofer, "Periodic solutions with prescribed minimal period for convex autonomous Hamiltonian systems", Invent. Math. 81 (1985), p. 155-188.

[10] I. Ekeland and R. Temam, "Convex analysis and variational problems", North-Holland-Elsevier, 1976.

[11] P. Rabinowitz, "Periodic solution of Hamiltonian systems", Comm. PAM 31 (1978) p. 157-184.

[12] L. Tonelli, "Fondamenti di Calcolo delle variazioni", Zanicheli, 1921-23 (and in "Opere Scelte", vol 3, Edizioni Cremonese, Roma, 1961).

I. Ekeland,
Centre de Recherche de
Mathématiques de la Decision,
Université de Paris IX,
Paris,
France.

M GIAQUINTA

Regularity of the minima in the calculus of variations

One of the best achievements of Tonelli was to introduce in a consistent and uniform way, the direct methods in the calculus of variations. As is well known now, the idea is to handle the problem of the existence of a minimum of an integral functional (bounded below)

$$F[u;\Omega] = \int_\Omega F(x,u,Du)dx \qquad (1)$$

in a class K of admissible functions u by introducing a convergence in K so that K itself is sequentially compact, and F is sequentially lower semi-continuous.

For the semicontinuity of F we need a rich topology, while for compactness the topology should not be so rich: the point is to realize a suitable compromise.

Tonelli worked in the class of absolutely continuous functions with uniform convergence, as they concerned unidimensional variational integrals. With the introduction of Sobolev spaces, the methods of Tonelli were soon extended by Morrey to multiple integrals of the calculus of variations.

Because of the contributions of many authors (Tonelli, Haar, Morrey, Serrin, De Giorgi, Ioffe, Cesari, Acerbi-Fusco ...) we can now state the

<u>Theorem</u> (lower semicontinuity). Let Ω be a bounded open subset of $\mathbb{R}^n$ and let $F(x,u,p) : \Omega \times \mathbb{R}^N \times \mathbb{R}^{nN} \to \mathbb{R}$ be a non-negative Carathéodory function, i.e. measurable in x and continuous in (u,p), satisfying the condition

$$0 \leq F(x,u,p) \leq c_1|p|^m. \qquad (2)$$

Then the functional

$$F[u;\Omega] = \int_\Omega F(x,u,Du)dx,$$

where $D_u = \{D_\alpha u^i\}$ is sequentially lower semicontinuous with respect to weak convergence in $H^{1,m}_{loc}(\Omega,\mathbb{R}^N)$ if and only if the integrand is quasi-convex in the sense of Morrey, that is, for each x_0, u_0, p_0 and $\phi \in C^\infty_0(\Omega,\mathbb{R}^N)$, we have

$$\int_\Omega [F(x_o,u_o,p_o + D\phi) - F(x_o,u_o,p_o)]dx \geq 0.$$

or, in other words, the linear (affine) functions are minima for the functional

$$\int_\Omega F(x_o,u_o,Dv)dx.$$

Let us note that the condition (3) of quasi-convexity is a global one, not local. However, if F is of class C^2, then condition (3) implies the local property

$$F_{p^i_\alpha p^j_\beta}(x,u,p)\xi^\alpha\xi^\beta\eta_i\eta_j \geq 0, \quad \forall\xi \in \mathbb{R}^n, \quad \forall\eta \in \mathbb{R}^N, \tag{4}$$

which is known as the Legendre-Hadamard condition. In turn this condition reduces to the hypothesis of convexity of $F(x,u,p)$ with respect to p in the scalar case $N = 1$. Indeed, in this case (4) reduces to

$$F_{p^i_\alpha p^j_\beta}(x,u,p)\ \xi^\alpha_i\xi^\beta_j \geq 0, \quad \forall\xi \in \mathbb{R}^{nN}. \tag{5}$$

The addition to condition (2) of a coercivity condition, typical in Tonelli's work,

$$F(x,u,p) \geq c_o|p|^m, \quad m > 1, \quad c_o > 0, \tag{6}$$

leads immediately, also in view of the reflexivity of the Sobolev spaces $H^{1,m}$, to the existence of a point of minimum, for instance, in the class of functions $v \in H^{1,m}(\Omega,\mathbb{R}^N)$ with given boundary values.

Thus, we can refer to (local) points of minimum u of the functional $F[v,\Omega]$ as those functions $u \in H^{1,m}_{loc}(\Omega,\mathbb{R}^N)$ such that for each ϕ with supp $\phi \subset\subset \Omega$ we have

$$F[u;\text{supp}\ \phi] \leq F[u + \phi;\ \text{supp}\ \phi].$$

And now the regularity problem presents itself in a natural way, or the 19th Hilbert problem: how regular is u?

The classical approach to this problem is via the Euler-Lagrange equation, and the variational equation. A first result, obtained after a research span of 50 years, is the following one: if the integrand F is C^∞, or analytic, and is regular, in the sense that it satisfies the strict Legendre-Hadamard condition:

$$F_{p^i_\alpha p^j_\beta}(x,u,p)\xi^\alpha\xi^\beta\eta_i\eta_j > c(M)|\xi|^2|\eta|^2 \quad \forall\xi \in \mathbb{R}^n, \quad \forall\eta \in \mathbb{R}^N,$$

for $|u| + |p| \leq M$, with $c(M) > 0$, then the points of minimum of class C' are C^ω, or analytic.

The regularity problem, therefore, reduces to the passage from $H^{1,m}$ to C^1. Again the main route is the Euler-Lagrange equation and the variational equation. But, this time, we are compelled to use supplementary assumptions, often not natural. In particular, we may have to require growth condition not only on the integrand function, but even on its deriatives, often we have to acquire the boundedness of the point of minimum, and in the vectorial case, we may have to require that the point of minimum be in modulus sufficiently small. For a discussion of the results, and of the difficulties that this approach has entitled, we refer, e.g. to [3]. We only note that these difficulties derive from the fact that this approach does not separate points of minimum, stationary points, or even solutions of nonlinear elliptic problems which are not the Euler-Lagrange equations of a functional.

In the present paper, I wish to present two results which can be seen as the counterpart in terms of regularity of the direct methods in the calculus of variations.

The models for these two results can be traced to two theorems, one due to Morrey, 1938 [8] (see also [9]), and one due to Tonelli 1920 [10] (see also [1]).

<u>Theorem 1</u> (Morrey) Let Ω be a bounded open set in $\mathbb{R}^2$ and $F(x,u,p)$ be a Carathéodory function satisfying

$$c_0|p|^2 \leq F(x,u,p) \leq c_1|p|^2.$$

Then the points of minimum $H^{1,2}_{loc}(\Omega,\mathbb{R}^N)$ of the functional

$$F[u,\Omega] = \int_\Omega F(x,u,Du)dx$$

are locally Hölder.

We note that no assumption of convexity or ellipticity is made in this theorem.

Theorem 2 (Tonelli). Assume $n = 1$, $N \geq 1$. Let u be a point of minimum for the functional

$$\int_I F(x,u,u')dx, \quad I = \text{interval in } \mathbb{R},$$

where $F(x,u,p)$ is a non-negative C^∞ function satisfying

$$F_{pp}(x,u,p) > 0.$$

Then there is a closed set Σ with meas $\Sigma = 0$ such that $u \in C^\infty(I-\Sigma)$. Moreover,

$$\lim_{\substack{\text{dist }(x,\Sigma)\to 0 \\ x \notin \Sigma}} |u'(x)| = +\infty,$$

and for each $x \in I$

$$u'(x) = \lim_{\delta\to 0} \frac{u(x+\delta)-u(x)}{\delta}$$

exists.

We can think of the following theorem as an extension and stronger version of Theorem 1, and at the same time as a description of regularity properties of points of minimum.

Theorem 3 (Giaquinta-Giusti [4]). Let $u \in H^{1,m}_{loc}(\Omega,\mathbb{R}^N)$ be a point of minimum for the functional

$$F[u;\Omega] = \int_\Omega F(x,u,Du)dx,$$

where F is a Carathéodory function satisfying

$$c_0|p|^m \leq F(x,u,p) \leq c_1|p|^m, \quad m > 1.$$

Then

(i) in the scalar case, $N=1$, u is locally Hölder in Ω with Hölder exponent which depends only on c_0, c_1, n, m.

(ii) in the vectorial case, $N \geq 1$, there is an exponent $r > m$, which depends only on c_0, c_1, n, m, such that $u \in H^{1,r}_{loc}(\Omega,\mathbb{R}^N)$; moreover, the gradient of u verifies a Hölder inequality "in reverse" with growing supports, i.e., for each ball $B_R \subset\subset \Omega$, we have

$$\left(\int_{B_{R/2}} |Du|^r dz\right)^{1/r} \leq c(\delta,c_1,n,m)\left(\int_{B_R} |Du|^q\, du\right)^{1/q} \qquad (*)$$

for each q with $0 < q \leq m$. In particular, if $m = n$, u is locally Hölder.

We refer to [3] and [4] for more details. Here we limit ourselves to the remark that, as examples of De Giorgi, Giusti-Miranda, Necas, Mazja (cf. e.g., [3]) show, the points of minimum in the vectorial case may be non-continuous, even if the integrand F is regular. It may even occur, in the hypotheses of Theorem 3, as an example of J. Soucek shows, (cf., e.g. [3]) that the points of minima may be singular for $n > 3$ in a dense set.

The following theorem, though less sharp than Theorem 2 in the description of the behaviour of the points of minimum, can well be considered, however, as an extension of Theorem 2.

Theorem 4 (Giaquinta-Modica [7]) Assume

(i) $c_0|p|^n \leq F(x,u,p) \leq c_1|p|^m$, $m \geq 2$;

(ii) $F(x,u,p)$ is of class C^2 in p and

$$|F_{pp}(x,u,p)| \leq c_2(1 + |p|^2)^{\frac{m-2}{2}};$$

(iii) $(1 + p^2)^{-m/2} F(x,u,p)$ is Hölder in (x,u) uniformly with respect to p;

(iv) F is strictly quasi-convex, i.e., for each x_0, u_0, p_0 and for each $\phi \in C_0^\infty(\Omega,\mathbb{R}^N)$

$$\int_\Omega [F(x_0,u_0,p_0 + D\phi) - F(x_0,u_0,p_0)]dx$$

$$\geq \nu\int_\Omega (1 + p_0^2 + |D\phi|^2)^{\frac{n-2}{2}} |D\phi|^2 dx, \quad \nu > 0.$$

(*) $\displaystyle ⨍_E f\, dx = \frac{1}{|E|}\int_E f\, dx.$

Let $u \in H^{1,m}_{loc}(\Omega,\mathbb{R}^N)$ be a point of minimum for

$$\int_\Omega F(x,u,Du)dx.$$

Then there is an open set $\Omega_o \subset \Omega$ such that $u \in C^{1,\mu}(\Omega_o,\mathbb{R}^N)$, and moreover means $(\Omega-\Omega_o) = 0$.

Theorem 4 is the final result of many stages of research. It was proved by Morrey, Giusti-Miranda, Giusti 1968 for $m \geq 2$ and $F = F(p)$; by Pepe for $1 < m < p$; by Giaquinta-Giusti in 1983 and Ivert in 1983; by Giaquinta-Ivert in 1984 for $m \geq 2$, $F = F(x,u,p)$. For a discussion of these results see e.g., [3], [5], [6]. Under assumptions of strict convexity the theorem for $F = F(p)$ was first proved by Evans [2].

Let us note that the regularity problems are still open under the hypotheses of Theorem 4 in the case $1 < m < 2$, and moreover many of the questions concerning the singular set are still open. We refer to [3] for a discussion and results.

In the scalar case, obviously, one expects regularity everywhere but this has been proved under the hypotheses of Theorem 4, only for $m = 2$.

Theorem 5 (Giaquinta-Giusti) Under the hypotheses of Theorem 4, if $N = 1$ and $m = 2$, then $\Omega_o = \Omega$.

REFERENCES

[1] L. Cesari, Optimization-Theory and Applications. Problems with Ordinary Differential Equations. Applications of Math. n° 17, Springer Verlag 1983.

[2] C.L. Evans, Quasiconvexity and partial regularity in the Calculus of Variations, preprint 1984.

[3] M. Giaquinta, Multiple integrals in the Calculus of Variations and nonlinear elliptic systems. Annals Math. Studies n° 105, Princeton Univ. Press, 1983.

[4] M. Giaquinta, E. Giusti, On the regularity of the minima of variational integrals. Acta Math. 148 (1982) 31-46.

[5] M. Giaquinta, E. Giusti, Differentiability of minima of non differentiable functionals. Inventiones Math. 72 (1983) 285-298.

[6] M. Giaquinta, P.A. Ivert, Partial regularity for minima of variational integrals. Arkiv för Math.

[7] M. Giaquinta, G. Modica, Partial regularity of minimizers of quasi-convex integrals. Ann. Inst. H. Poincaré, Analyse non linéaire.

[8] C.B. Morrey Jr., Existence and differentiability theorems for the solutions of variational problems for multiple integrals. Bull. Amer. Math. Soc. 46 (1940) 439-458.

[9] C.B. Morrey Jr., Multiple integrals in the Calculus of Variations. Springer Verlag 1966.

[10] L. Tonelli, Fondamenti di Calcolo delle Variazioni. vol. 2, Zanichelli, Bologna 1921-23.

M. Giaquinta
Istituto di Matematica Applicata
Facoltà Ingegneria
50134 Firenze
Italy.

S HILDEBRANDT

Remarks on some isoperimetric problems

At this meeting, we celebrate the hundredth birthday of Leonida Tonelli who has contributed so many beautiful and profound results to the calculus of variations. I should in particular like to mention his solution of Plateau's problem [26] that considerably simplified the work of J. Douglas and T. Radó. Tonelli's new approach to Plateau's problem was, independently and in the same year, found by R. Courant [3].

In this paper, I shall report on an isoperimetric problem that is closely related to the aforementioned work of Tonelli and Courant. It leads to a free boundary problem for surfaces of constant mean curvature.

The present notes basically yield a survey of some results for the so-called *partition problem* which were found by Grüter, Nitsche, and the author in the joint work [12].

1. A PARTITION PROBLEM. BOUNDARY REGULARITY

Let K be a bounded convex body in $\mathbb{R}^3$ with boundary T. We consider the following *partitioning problem* P :

Determine a rectifiable surface $\bar{S}$ *of minimal or stationary area, with boundary* Σ *contained in* T, *which divides* K *into two parts* K_1 *and* K_2 *such that*

$$\text{meas } K_1 = \sigma \text{ meas } K, \quad \text{meas } K_2 = (1-\sigma)\text{meas } K,$$

where σ *denotes a preassigned constant with* $0 < \sigma < 1$.

Bokowski and Sperner jr. [2] have proved the existence of minimal partitions of the ball using tools from geometric measure theory, and Almgren [1] has established existence and regularity almost everywhere of minimal partitions of $\mathbb{R}^n$. Applying the work of De Giorgi, one can obtain sets of finite perimeter as minimal solutions of the afore stated partition problem, and interior regularity follows from results by Gonzalez, Massari, and Tamanini [8], and Giusti [7].

There are, however, no existence results for partitioning surfaces that are stationary and have a prescribed topological type, say, the type of a disk.

Boundary regularity for minimizing partitioning surfaces was first verified by Hildebrandt and Wente [18], and for stationary surfaces, it was established in [12]. We shall state this result below.

One first realizes that every solution $\bar{S}$ of P is a surface of constant mean curvature (in special situations: a minimal surface), perpendicular to the boundary T. Interior regularity follows from Grüter's work [9, 10]. Thus the emphasis of [12] lies in the proof of boundary regularity for $\bar{S}$, and it will already be assumed that the interior part S of $\bar{S}$ is a regular C^1-manifold which divides int K - S into two disjoint parts of preassigned measure. As a consequence, S has a conformal parameter representation on some Riemann surface B. Generally speaking, the surface S need not be of disk-type. It could, for instance, also have the topological type of an annulus, as is the case for a cylinder, a catenoid, or an arbitrary Delaunay surface. Although the approach of [12] can easily be modified to handle this case or more general cases as well, we shall restrict ourselves here to solutions of P which can be parametrized over a disk.

In order to have a clear cut situation, we shall fix the assumptions which will be used during the proof of the subsequent boundary regularity theorem.

Assumption (A1)

(i) S has a conformal parameter representation $x : B \to \mathbb{R}^3$ of class $C^1(B,\mathbb{R}^3)$ on the unit disk B (that is, (1.2) holds).

(ii) S has finite area, so that its representation x is of class $H_2^1(B,\mathbb{R}^3)$. We assume that $\Sigma := x(\partial B)$ is contained in T, where the boundary condition $\Sigma \subset T$ is to be interpreted in the sense that the L^2-trace of x on ∂B lies in T.

(iii) The surface $S = x(B)$ omits a neighbourhood of some point $p \in T$.

(iv) The map $x : B \to \mathbb{R}^3$ is an embedding of B into the interior of K such that

$$\text{int } K - S = \Omega_1 \cup \Omega_2, \quad \Omega_1 \cap \Omega_2 = \emptyset,$$

where Ω_1 and Ω_2 are simply-connected regions and

$$\text{meas } \Omega_1 = \sigma \text{ meas } K, \quad \text{meas } \Omega_2 = (1 - \sigma)\text{meas } K.$$

The main result of [12] then is the following

<u>Theorem 1</u> *Let* T *be a regular surface of class* C^3, *or* $C^{m,\alpha}(m \geq 3, 0 < \alpha < 1)$, *or* C^ω, *respectively. Suppose that* $S = \{x(w) \in \mathbb{R}^3 : w \in B\}$ *satisfies assumption* (A1) *as well as the following variational property* (A2): S *is stationary for the area functional (Dirichlet integral) in the class of all disk-type surfaces with boundary on* K *which partition* K *into two parts of measures* σ meas K and $(1 - \sigma)$meas K.

Then $x(w)$ *is a real-analytic surface of constant mean curvature* H *on* B. *Moreover,* $x(w)$ *is of class* $C^{1,\beta}(\bar{B},\mathbb{R}^3)$ *for every* $\beta \in (0,1)$, *or of class* $C^{m,\alpha}(\bar{B},\mathbb{R}^3)$, *or real-analytic on* $\bar{B}$, *respectively, and* $\bar{S} = S \cup \Sigma$ *intersects* T *orthogonally in the points of* $\Sigma = x(\partial B)$.

The mapping $x(w)$ *satisfies the equation*

$$\Delta x = 2H\, x_u \wedge x_v \tag{1.1}$$

as well as the conformality relations

$$|x_u| = |x_v|, \quad x_u \cdot x_v = 0 \tag{1.2}$$

in B, *and it intersects* T *perpendicularly.*

As it stands, assumption (A2) lacks precision. This precision will be supplied below.

It should be noted that the convexity of the surface T is not essential for the proof of Theorem 1. It is mainly used to obtain a simple form of assumption (A1).

We note that surfaces of different mean curvature may be solutions of the same partitioning problem. If, for instance, we want to decompose a regular tetrahedron into two parts of equal measure, then one stationary solution is formed by an isosceles triangle Δ that passes through one of the corners of the tetrahedron and meets the opposite face perpendicularly. Another stationary solution is given by a spherical surface S of suitable radius R which is centred at one of the corners of the tetrahedron. It, moreover, turns out that S has less area than Δ, and clearly Δ is a minimal surface $(H = 0)$, whereas S has mean curvature $H = 1/R$. The spherical surface S is a relative minimum for the partitioning problem whereas the plane surface Δ does not furnish a minimum since, in every neighbourhood of Δ, there is an admissible surface with smaller area than Δ. By smoothing the corners and edges of the tetrahedron in a suitable way, we can even generate a smoothly bounded convex body, arbitrarily close to the tetrahedron, which is partitioned (into two parts of equal measure) by a spherical surface S and by a plane surface Δ where Δ is only stationary while S yields a relative minimum.

Another instructive example is provided by a box K (that is, a solid rectangle). Denote by a,b,c the length of the edges of K, and suppose that $a > b > c$. Then we have three continua of minimal surfaces of different area ab, ac, bc, which consist of planar cross-sections of the box parallel to its faces. For a given value of $\sigma \in (0,1)$, each of these continua contains exactly one minimal surface that partitions the box in the prescribed way and, actually, is a relative minimum of the partitioning problem.

Other relative minima are formed by hemispheres sitting on of the faces of the box. In fact, they yield 2-parameter families of solutions to the same value of σ. Further solutions are given by circular cylinders which perpendicularly meet two parallel faces of the box. These cylinder surfaces become unstable if their length exceeds half of their parameter. On the other hand, if R is defined by $\frac{2}{3}\pi R^3 = \sigma \cdot \text{meas } K$, then the surface area of a hemisphere of radius R is larger than the area of a cylinder of length c enclosing the same volume as the hemisphere, provided that $R < (2/3)c$. Still the cylinder does not furnish the absolute minimum of area since suitable hemicylinders, quarter cylinders or spherical caps yield a smaller value for the area of the partitioning surface while, together with the boundary of the box, they enclose the same volume.

Other stationary solutions are suitable pieces of unduloids, that is, of surfaces of revolution, the generating curves of which are "roulades" of foci of ellipses, when these ellipses role on a straight line. There exist also stationary solutions with "many arms". For instance, the adjoint to the minimal surface within a regular quadrilateral, by reflection, generates a celebrated periodic minimal surface discovered by H.A. Schwarz. An appropriate part of this periodic surface meets each face of a certain cube at a right angle, and thus it forms a stationary solution of the partitioning problem, with six arms reaching to the boundary of the cube.

By reflecting this cube sufficiently often, we may build a box and a minimal surface S which meets the boundary of the box orthogonally with n arms and has the genus g. We can obviously achieve that n and g become larger than any given number. Employing a scaling argument, we see that there is a number $\sigma \in (0,1)$ such that the σ-partitioning problem for some given cube possesses infinitely many stationary solutions of different topological type, and these solutions are minimal surfaces. Intuitively, it seems plausible that for nearby values of σ the partitioning problem again has infinitely many solutions which form surfaces of constant mean curvature $H \neq 0$. It would suffice to prove that for values of H close to zero, there exist embedded surfaces of mean curvature H within a cube which orthogonally attach six arms to the six faces of the cube. Similarly to Schwarz's periodic minimal surface, we then could construct periodic H-surfaces with a cubic lattice, by employing the following reflection principle which is an immediate consequence of Theorem 1:

REFLECTION PRINCIPLE *If a surface of constant mean curvature* H *meets a plane* E *at a right angle, then, by reflection at* E, *it can be extended as a real analytic surface of mean curvature* H.

We however note that extension by reflection at straight lines is valid only for minimal surfaces but not for H-surfaces with $H \neq 0$.

The reflection principle for H-surfaces has also been stated in [23].

Now we shall make assumption (A2) precise. To this end, we suppose that the surface $S = \{x(w) : w \in B\}$ satisfies assumption (A1) and, furthermore, that S does not meet an $\mathbb{R}^3$-neighbourhood $U(p)$ of some point $p \in T$. Let $U_1(p)$ be some smaller neighbourhood of p with $U_1(p) \subset\subset U(p)$. Then we shall construct a vector field $Q(x) = (Q^1(x), Q^2(x), Q^3(x))$ on $\mathbb{R}^3$ such that the following holds:

(i) Q is of class C^1 on $\mathbb{R}^3$ and satisfies
$\operatorname{div} Q = 1$ in K.

(ii) $Q|_T$ is a tangential vector field on $T - U_1(p)$, that is, the normal component $Q_n = n \cdot Q$ of Q satisfies $Q_n(x) = 0$ for all $x \in T - U_1(p)$. (Here n denotes the exterior normal of $T = \partial K$.)

Such a vector field Q can be obtained in the special form $Q(x) = \operatorname{grad} f(x)$, where the scalar function f(x) satisfies

$$\Delta f = 1 \text{ in } K$$

$$\frac{\partial f}{\partial n} + \gamma f = 0 \text{ on } T = \partial K,$$

and γ stands for some sufficiently regular function on T which satisfies $\gamma \geqq 0$, $\gamma \not\equiv 0$, and finally $\gamma \equiv 0$ on $T - U_1(p)$ (cf. [6], p. 124).

Now let $S = \{x = x(w) : w \in B\}$, be a solution of the partition problem. Then we infer from (i) and (ii) that, for the component Ω_1 of int $\{K - S\}$ which appears in (A1), we have

$$\text{meas } \Omega_1 = \int\int\int_{\Omega_1} dx^1 dx^2 dx^3 = \int\int\int_{\Omega_1} \text{div } Q \, dx^1 dx^2 dx^3$$

$$= \int\int_{\partial\Omega_1} Q_n dA = \iint_B Q(x) \cdot x_u \wedge x_v du \, dv.$$

That is

$$\int\int_B Q(x) \cdot x_u \wedge x_v du \, dv = \text{meas } \Omega_1. \quad (1.3)$$

The precise formulation of (A2) then reads as follows:

Assumption (A2) The surface $x(w)$ is stationary for the Dirichlet integral $D[z] = \frac{1}{2}\iint_B |\nabla z|^2 du \, dv$ within the class C^* of surfaces $z \in C(T)$ which satisfy

$$V^Q[z] := \int\int_B Q(z) \cdot z_u \wedge z_v du \, dv = \sigma \cdot \text{meas } K =: c. \quad (1.4)$$

Here, the set $C(T)$ is defined by

$$C(T) = \{z : z \in H_2^1(B, \mathbb{R}^3), \quad z(\partial B) \subset T\},$$

and $x(w)$ is called stationary within C^* if

$$\lim_{t \to 0} \frac{1}{t} \{D[x_t] - D[x]\} = 0$$

holds for every variation x_t of x which satisfies

$$V^Q[x_t] = c$$

and is of one of the two following types:

TYPE 1. The surfaces x_t, $|t| < t_0$, are of the form $x_t = x \circ \tau_t$, where $\{\tau_t\}_{|t|<t_0}$ is a family of diffeomorphisms from $\bar{B}$ to itself such that τ_0 is the identity, and that $\tau(t,w) := \tau_t(w)$ is of class C^1 on $(-t_0,t_0) \times \bar{B}$.

TYPE 2. The surfaces x_t are of the form

$$x_t(w) = x(w) + t\Psi(w,t), \quad |t| < t_0,$$

where

$$D[\Psi(\cdot,t)] \leq C \text{ for } |t| < t_0$$

with a bound C independent of t, and

$$\lim_{t\to 0} \Psi(w,t) = \Phi(w) \quad \text{a.e. on } B$$

for some $\Phi \in H_2^1(B,\mathbb{R}^3)$.

The gist of this approach is that we have reduced the partition problem P to a free boundary problem for the Dirichlet integral with the subsidiary condition $V^Q[x] = c$. Using ideas from [18], one can establish the existence of a Lagrange parameter μ, and thus P is in fact reduced to a free boundary problem for the functional $D[x] + \mu V^Q[x]$ in the class $C(T)$, without any further subsidiary condition. The boundary regularity for solutions of this problem can, however, be proved by a suitable extension of the methods developed in [11]. For details we refer the reader to [12].

2. LENGTH AND AREA ESTIMATES FOR SOLUTIONS OF THE PARTITION PROBLEM

The next result is also taken from [12].

Theorem 2 *Let* K *be a convex body which contains a ball of radius* R_0 *and has a smooth boundary* T. *Suppose also that* $R > 0$ *is a lower bound for the principal curvature radii of* T. *Now let* $S = \{x(w) : w \in B\}$ *be a stationary, disk-type solution of a partition problem* P *for* K *which has the free trace* $\Sigma = \{x(w) : w \in \partial B\}$ *on* T *and the (constant) mean curvature* H. *Denote by* $L(\Sigma) = \int_{\partial B}|dx|$ *the length of* Σ, *and by* $A(S) = D[x]$ *the area of* S. *Then we have the lower bound*

$$L(\Sigma) \geqq \frac{2\pi R_o}{1+(\text{diam } K-R_o)\ |H|} \tag{2.1}$$

for $L(\Sigma)$ *as well as the upper bound*

$$L(\Sigma) \leqq 2(\frac{1}{R} + |H|)A(S) \tag{2.2}$$

in terms of S. *Moreover,* S *has only finitely branch points. These branch points are of finite order* m_j , *and*

$$2\ (1 + \sum_j m_j) \leqq \int_\Sigma \kappa ds + H^2 A(S) \tag{2.3}$$

where κ *denotes the curvature of the trace* Σ.

Proof We introduce polar coordinates ρ,θ on B by $w = u + iv = \rho e^{i\theta}$ and write $x(\rho,\theta)$ instead of $x(w)$. Then Green's formula

$$\int_{\partial B} \phi \cdot x_\rho d\theta = \iint_B \phi \cdot \Delta x\ du\ dv + \iint_B \nabla\phi \cdot \nabla x\ du\ dv \tag{2.4}$$

holds for all $\phi \in C^1(\bar{B},\mathbb{R}^3)$. As before, let $n(x)$ be the exterior unit normal field on T, and denote by $s = s(\theta)$ the parameter of arc length on Σ. We set $\xi(s) = x(1,\theta(s))$. Then(*) $x_\rho(1,\theta) = |x_\rho(1,\theta)| \cdot n(x(1,\theta))$, since (1.2) holds and x meets T perpendicularly at Σ. Thus, by (1.1),

$$2H \iint_B x_u \wedge x_v du\ dv = \int_\Sigma n(\xi(s))ds, \tag{2.5}$$

$$\int_\Sigma \phi \cdot n(\xi(s))ds = 2H\int_B\int \phi \cdot (x_u \wedge x_v)du\ dv + \int_B\int \nabla\phi \cdot \nabla x\ du\ dv. \tag{2.6}$$

Since

$$2\int_B\int x_u \wedge x_v du\ dv = \int_{\partial B} x \wedge dx = \int_\Sigma \xi \wedge d\xi,$$

(*) We also note that, because of the convexity of T and $\bar{S} \subset K$, there exist no branch points of x on ∂B.

it follows from (2.5) that

$$\int_\Sigma \{n(\xi)ds - H\xi \wedge d\xi\} = 0. \tag{2.7}$$

Let $L = L(\Sigma)$, and $\bar{\xi} := ⨍_0^L \xi(s)ds = \frac{1}{L}\int_0^L \xi(s)ds$. Then Wirtinger's inequlity yields

$$\int_0^L |\xi - \bar{\xi}|^2 ds \leqq \frac{L^3}{4\pi^3}. \tag{2.8}$$

We can assume that the ball $\{z \in \mathbb{R}^3 : |z| \leqq \mathbb{R}_0\}$ is contained in K whence the support function $p(z) = z \cdot n(z)$ of the convex body K satisfies

$$p(z) \geqq R_0 \text{ for all } z \in T.$$

Thus

$$\begin{aligned} R_0 L - \int_0^L \bar{\xi} \cdot n(\xi)ds &\leqq \int_0^L (\xi - \bar{\xi}) \cdot n(\xi)ds \\ &\leqq \sqrt{L} \cdot \{\int_0^L |\xi - \bar{\xi}|^2 ds\} \leqq \frac{L^2}{2\pi}. \end{aligned} \tag{2.9}$$

Set $(a,b,c) = a \cdot (b \wedge c)$ for $a,b,c \in \mathbb{R}^3$. Then

$$(\bar{\xi},\xi,\xi') = (\bar{\xi},\xi - \bar{\xi},\xi'),$$

and we can obtain from (2.7), (2.8), and (2.9) that

$$\begin{aligned} R_0 L &\leqq \frac{L^2}{2\pi} + H\int_0^L (\bar{\xi},\xi - \bar{\xi},\xi')ds \\ &\leqq \frac{L^2}{2\pi} + |H| \int_0^L |\bar{\xi}|\ |\xi - \bar{\xi}||\xi'|ds \\ &\leqq \frac{L^2}{2\pi} + |H|\ |\bar{\xi}|\ \{\int_0^L |\xi - \bar{\xi}|^2 ds\}^{1/2} \sqrt{L} \\ &\leqq \frac{L^2}{2\pi}(1 + |H|\ |\bar{\xi}|). \end{aligned}$$

Obviously, $|\bar{\xi}| \leqq \text{diam } K - R_0$, and (2.1) is proven.

Estimate (2.2) is obtained as follows. We choose (as in [20]) a vector field $\eta(x)$ which coincides with the normal field $n(x)$ on T and vanishes outside a suitable parallel strip of T. Then, for every $\varepsilon > 0$, we can construct $\eta(x)$ in such a way that

$$|\eta(x)| \leqq 1, \quad |\text{grad } \eta(x)| \leqq \frac{1+\varepsilon}{R} .$$

Using $\phi(w) = \eta(x(w))$ in (2.6), and observing that

$$|\nabla\phi \cdot \nabla x| \leq |\text{grad } \phi(x)| \; |\nabla x|^2 \leqq 2 \cdot \frac{1+\varepsilon}{R} \cdot \frac{1}{2}|\nabla x|^2,$$

$$|x_u \wedge x_v| = \frac{1}{2}|\nabla x|^2,$$

we arrive at

$$L(\Sigma) \leqq 2(\frac{1+\varepsilon}{R} + |H|)D[x].$$

Then (2.2) follows, as $\varepsilon \to 0$.

The estimate (2.3) is a consequence of the Gauss-Bonnet theorem (cf. [13]).

Remark 1 The estimates (2.1-3) still involve the unknown mean curvature H which - in an unknown way - depends on the parameter σ appearing in the partition problem. Thus it will be useful to know an estimate of $|H|$ from above. Such an estimate can be obtained as follows: from (1.1), we get

$$\iint_B Q(x) \cdot \Delta x \, du \, dv = 2H \, V^Q[x]$$

where $Q(x)$ is the vector field determined at the end of Section 1. Since $n \cdot Q = 0$ on T, we conclude that $Q(x) \cdot x_\rho = 0$ on ∂B. Thus (2.4) implies

$$2H \, V^Q[x] = -\iint_B \nabla Q(x) \cdot \nabla x \, du \, dv \qquad (2.10)$$

where $\nabla = (\frac{\partial}{\partial u}, \frac{\partial}{\partial v})$. Since

$$|\nabla Q(x) \cdot \nabla x| \leq |\text{grad } Q(x)| \quad |\nabla x|^2 \leqq 2|Q|_{1,K} \cdot \frac{1}{2}|\nabla x|^2,$$

and since equation (1.4) of (A2) yields

$$V^Q[x] = \sigma \cdot \text{meas } K,$$

we obtain

$$|H| \leq \frac{|Q|_{1,K} \cdot A(S)}{\sigma \cdot \text{meas } K} \quad . \tag{2.11}$$

It seems to be a difficult task to estimate the area A(S) in terms of geometric properties of K and by the value of σ. In fact, no estimate will be possible if no further conditions are imposed on S. For instance, the planar minimal surface $S = \{x(w) : w \in B\}$, given by

$$x(w) = (\text{Re } w^m, \text{ Im } w^m, 0), \quad m \in \mathbb{N},$$

is stationary for the partition problem of the unit ball $\{z \in \mathbb{R}^3 : |z| \leq 1\}$ with $\sigma = 1/2$, and

$$A(S) = m\pi.$$

In other words, the area of S tends to infinity with the order m of the false branch point w = 0. However, it might be possible to estimate A(S) from above if S is assumed to be an embedded surface (as we have done in Section 1). It is trivial to get an upper bound of A(S) for *minimal* partitioning surfaces. One can, for instance, slice K by a family $\{E_t\}_{t \in \mathbb{R}}$ of parallel planes E_t. Then

$$\alpha := \sup_{t \in \mathbb{R}} A(E_t \cap K) \tag{2.12}$$

is an upper bound of A(S) for all minimal partitioning surfaces S, and (2.11) yields

$$|H| \leq \frac{\alpha \cdot |Q|_{1,K}}{\sigma \cdot \text{meas } K} \,. \tag{2.13}$$

This reasoning is too crude if we want to show that $|H| \to \infty$ for a minimal partition S as $\sigma \to 0$. However, for small values of σ, say

$$\sigma = \frac{1}{\text{meas } K} \cdot \frac{2\pi\varepsilon^3}{3}, \; \varepsilon \to +0, \tag{2.14}$$

surfaces S_ε close to hemispheres of radius ε will be good comparison surfaces, with

$$A(S_\varepsilon) = 2\pi\varepsilon^2[1 + O(1)] \text{ as } \varepsilon \to +0.$$

Thus (2.11) implies that the mean curvature H of a minimal partitioning surface S with a partition parameter σ determined by (2.14) satisfies

$$|H| \leq 3|Q|_{1,K} \cdot \varepsilon^{-1}\{1 + O(1)\} \text{ as } \varepsilon \to +0,$$

or

$$|H| \leq \text{const} \cdot \sigma^{-1/3} \text{ as } \sigma \to +0. \tag{2.15}$$

In fact, one expects that $|H| \sim \sigma^{-1/3}$ as $\sigma \to +0$.

Remark 2 If $S = \{x(w) : w \in B\}$ is a minimal surface (i.e., $H = 0$) which intersects T orthogonally, then (2.1) reduces to

$$2\pi R_0 \leq L(\Sigma). \tag{2.16}$$

We also obtain a bound of A(S) from below. For this purpose, we first notice that

$$\begin{aligned} R_0 L &\leq \int_0^L \xi \cdot n(\xi) ds = \int_{\partial B} x \cdot x_\rho \, d\theta \\ &= \iint_B |\nabla x|^2 du\, dv = 2A(S). \end{aligned} \tag{2.17}$$

Secondly, the isoperimetric inequality yields

$$A(S) \leq \frac{1}{4\pi} L^2. \tag{2.18}$$

Therefore

$$R_0^2 L^2 \leq 4A(S)A(S) \leq 4A(S) \cdot \frac{1}{4\pi} L^2 = \frac{L^2}{\pi} A(S),$$

whence

$$\pi R_0^2 \leq A(S). \tag{2.19}$$

On the other hand, the estimates

$$R_0 L \leq 2A(S) \leq \frac{1}{2\pi} L^2$$

yield once again (2.16). Nevertheless we now have proved a stronger result. Namely, formula (2.1) was derived under the assumption that S is of the type of a disk. But

$$R_0 L \leq 2A(S) \tag{2.17}$$

holds for all stationary minimal surfaces which are parametrized on an arbitrary planar domain, and the isoperimetric inequality (2.18) is true for minimal surfaces that are either of disk-type or of annulus type (cf. [22], [24], and Osserman-Schiffer [25]).

Thus (2.16) and (2.19) hold for all minimal surfaces S that intersect T orthogonally, and which are parametrized over a simply or doubly connected parameter domain.

REFERENCES

[1] Almgren, F.J., Existence and regularity almost everywhere of solutions to elliptic variational problems with constraints. *Mem. Amer. Math. Soc.*, No. 165, (1976).

[2] Bokowski, J. and Sperner, E., Zerlegung konvexer Körper durch minimale Trennflächen. *J. Reine Angew. Math.*, 311/312 (1979), 80-100.

[3] Courant, R., On the problem of Plateau. *Proc. Nat. Acad. Sci. USA* 22 (1936), 367ff, June 1936 (see also: *Ann. Math.* 38 (1937), 679-725).

[4] Dziuk, G., Uber die Stetigkeit teilweise freier Minimalflächen. *Manuscripta Math.*, 36 (1981), 241-251.

[5] Dziuk, G., *Über die Glattheit des freien Randes bei Minimalflächen.* Habilitationschrift, Aachen 1982.

[6] Gilbarg, D. and Trudinger, N.S., *Elliptic partial differential equations of second order.* Springer, Berlin-Heidelberg-New York 1977.

[7] Giusti, E., The equilibrium configuration of liquid drops. *J. Reine Angew. Math.* 321 (1981), 53-63.

[8] Gonzalez, E., Massari, U. and Tamanini, I., On the regularity of boundaries of sets minimizing perimeter with a volume constraint. *Indiana Univ. Math. J.*, 32 (1983), 25-37.

[9] Grüter, M., *Über die Regularität schwacher Lösungen des Systems* $\Delta x = 2H(x)x_u \wedge x_v$. Dissertation, Düsseldorf 1979.

[10] Grüter, M., Regularity of weak H-surfaces. *J. Reine Angew. Math.*, 329 (1981), 1-15.

[11] Grüter, M., Hildebrandt, S. and Nitsche, J.C.C., On the boundary behaviour of minimal surfaces with a free boundary which are not minima of the area. *Manuscripta Math.*, 35 (1981), 287-410.

[12] Grüter, M., Hildebrandt, S. and Nitsche, J.C.C., Regularity for surfaces of constant mean curvature with free boundaries. *Acta math.*, 156 (1986), 119-152.

[13] Heinz, E. and Hildebrandt, S., On the number of branch points of surfaces of bounded mean curvature. *J. Differential Geom.*, 4 (1970), 227-235.

[14] Hildebrandt, S. and Jäger, W., On the regularity of surfaces with prescribed mean curvature at a free boundary. *Math. Z.*, 118 (1970), 289-308.

[15] Hildebrandt, S. and Nitsche, J.C.C., Minimal surfaces with free boundaries. *Acta Math.*, 143 (1979), 251-272.

[16] Hildebrandt, S. and Nitsche, J.C.C., Optimal boundary regularity for minimal surfaces with a free boundary. *Manuscripta Math.*, 33 (1981), 357-364.

[17] Hildebrandt, S. and Nitsche, J.C.C. Geometric properties of minimal surfaces with free boundaries. *Math. Z.*, 184 (1983), 497-509.

[18] Hildebrandt, S. and Wente, H.C., Variational problems with obstacles and a volume constraint. *Math. Z.*, 135 (1973), 55-68.

[19] Hildebrandt, S. and Widman, K.-O., Some regularity results for quasi-linear elliptic systems of second order. *Math. Z.*, 142 (1975), 67-86.

[20] Küster, A., An optimal estimate of the free boundary of a minimal surface. *J. Reine Angew. Math.* 349 (1984), 55-62.

[21] Küster, A., *Zweidimensionale Variationsprobleme mit Hindernissen und völlig freien Randbedingungen.* Dissertation, Bonn 1983.

[22] Nitsche, J.C.C., *Vorlesungen über Minimalflächen.* Springer, Berlin-Heidelberg-New York 1975.

[23] Nitsche, J.C.C., Stationary partitioning of convex bodies. *Arch. Rational Mech. Anal.* 89 (1985), 1-19.

[24] Osserman, R., The isoperimetric inequality. *Bull. Amer. Math. Soc.* 84 (1978), 1182-1238.

[25] Osserman, R. and Schiffer, M., Doubly-connected minimal surfaces. *Arch. Rat. Mech. Anal.* 58 (1975), 285-307.

[26] Tonelli, L., Sul problema di Plateau, I & II. *Rend. R. Accad. dei Lincei* 24 (1936), 333-339, 393-398 (cf. also: Opere scelte, Vol. III, 328-341).

[27] Thompson, D'Arcy W., *On growth and form.* Cambridge University Press, abridged ed. 1969.

[28] Thompson, W., On the division of space with minimum partitional area. *Acta Math.*, 11 (1887/88), 121-134.

S. Hildebrandt
Mathematisches Institut
Universität Bonn
5300 Bonn
West Germany.

A MARINO

Evolution equations and multiplicity of critical points with respect to an obstacle

INTRODUCTION

I wish to give, in this paper, some of the ideas and the main results arising from research done in collaboration with E. De Giorgi and with M. Degiovanni, D. Scolozzi, M. Tosques.

(a) Our goal is to study a theoretical framework which is able to treat some problems in analysis and geometry which can be reduced to the study of "critical points" and "evolution equations" in a context where neither the usual regularity assumptions nor the convexity conditions, which are typical in monotone operator theory [5], are verified. A significant class of such problems arises, for instance, when some natural constraints are introduced in elliptic and parabolic equations.

Two typical cases are given by the following functionals.

(1) Let Ω be an open subset of $\mathbb{R}^n$, ϕ a real function on Ω, ρ a positive number and G a real function on $\Omega \times \mathbb{R}$. Set

$$\begin{cases} f_1(u) = \frac{1}{2}\int_\Omega |Du|^2\,dx + \int_\Omega G(x,u)dx & \text{if } u \in H_0^1(\Omega),\ u \leqq \phi, \\ & \int_\Omega u^2 dx = \rho^2 \\ f_1(u) = +\infty, & \text{elsewhere} \end{cases}$$

where $|Du|^2 = \Sigma_i (D_i u)^2$.

We can say that the set $\{u \mid \int_\Omega u^2\,dx = \rho^2,\ u \leqq \phi,\ u = 0 \text{ on } \partial\Omega\}$ is the "constraint" of the problem.

(2) Let Ω be an open subset of $\mathbb{R}^n$, G a real function on $\Omega \times \mathbb{R}^N$, and $\bar{u}$ a $\mathbb{R}^N$-valued function on $\partial\Omega$.

If $u = (u_1,\ldots,u_N)$, set

$$\begin{cases} f_2(u) = \frac{1}{2}\int_\Omega |Du|^2 dx + \int_\Omega G(x,u)dx, \text{ if } u \in H^1(\Omega,\mathbb{R}^N), \quad \Sigma_j u_j^2 \geq 1, \\ \qquad\qquad\qquad\qquad\qquad\qquad\qquad\qquad u = \bar{u} \text{ on } \partial\Omega \\ f_2(u) = +\infty, \quad \text{elsewhere} \end{cases}$$

where $|Du|^2 = \Sigma_{i,j}(D_i u_j)^2$.

We can say that the set $\{u|\Sigma_j u_j^2 \geq 1, u = \bar{u} \text{ on } \partial\Omega\}$ is the "constraint" of the problem.

In both cases one can raise either the problem of "critical points" of f_1 and f_2 (elliptic problem) or the problem of the "evolution equation" associated with these functionals (parabolic problem).

We remark that these functionals cannot be treated by means of the theory of regular perturbations of convex functions, as the set where they are finite is not convex.

On the contrary, if we drop out, in (1), the condition $\int_\Omega u^2 dx = \rho^2$, or we substitute, in (2), the $\Sigma_j u_j^2 \geq 1$ by $\Sigma_j u_j^2 \leq 1$, we get the classical elliptic and parabolic variational inequalities which have been brilliantly fitted in the theory of monotone operators mentioned above.

This is the reason why we believed it was important to develop our study starting from the framework of that theory, even if many other important studies have gone in different directions (see, for example, [10], [31]).

(b) It should be noted that, in the previous examples, the study of a functional f on a "constraint" is carried out by setting $f = +\infty$ outside the constraint, according to the technique which has been very useful in the theory of convex functions.

Actually the definitions of subdifferential, gradient (see (1.4)) and evolution equation (see (1.3) and (1.8)) give back the well known notions, when we study a regular function on a regular manifold. On the other hand, by this method, we can work with functionals defined on normed linear spaces.

This allows a compact treatment and suggests some meaningful formulations of constrained problems (see (1.5)).

(c) It is useful to point out that the fact alone that the functionals we study have a nonconvex domain (the set where they are finite) suggests that they may have several critical points: the number of such points can be evaluated by means of the "topological type" of the domain, extending to such functionals the techniques of Lusternik and Schnirelman [23], [33], which are essentially based on the associated evolution equation.

This line is followed to solve the problems presented in Sections 3 and 5, by means of the theories of "critical points" and "evolution equations" described in Sections 2 and 4.

The theorem which links the topological type of the domain of some lower semicontinuous functionals with the number of their "critical points" is given in (3.3).

(d) Schematically, we have studied the evolution equation

$$0 \in u' + A(u)$$

either in the case when A is the "subdifferential" of certain functionals (variational problem), or in the general case.

The variational problem has been considered either in some compactness assumptions (see Section 2) or in hypotheses which allow the inclusion of the theory of convex functions (see Section 4).

Both these cases allowed us to apply some interesting functionals, such as f_1 and f_2, to the study of "geodesics with respect to an obstacle" (existence and their number) and the related evolution equation (see Section 3) and to the study of the eigenvalues of the Laplace operator "with respect to an obstacle" (existence and their number) and the related evolution equation (see Section 5).

In this exposition, I shall refer only to variational problems.

The studies in the non variational case ((ϕ, f)-monotone operators [11], [21]) lead on statements which include the theory of monotone operators and which apply, for instance, also to some hyperbolic systems (see [14], [21]).

A non autonomous evolution equation in a variational case has been studied in [15].

An extension of these techniques to Volterra integral equations has been considered in [29].

Some results concerning the bifurcation problem, associated with the problem referred to in Section 5, can be found in [16]. Other results and studies are mentioned elsewhere in this exposition.

1. SLOPE, SUBDIFFERENTIALS AND CURVES OF MAXIMAL SLOPE

It has been necessary to consider suitable definitions which extend the classical ones of gradient and evolution equation. In [12], some definitions and existence theorems were introduced. These definitions were studied and compared in [27]; in this section I recall some of them, while I refer to the next section for existence theorems.

Consider a metric space (X,d) and a function $f : X \to \mathbb{R} \cup \{+\infty\}$.

The set $D(f) = \{u \in X \mid f(u) < +\infty\}$ is said to be the "domain of f".

(1.1) Definition

Let u be in D(f): we call the following number the "(descendent) slope of f at u"

$$|\nabla f|(u) = - \liminf_{v \to u} \frac{[f(v)-f(u)] \wedge 0}{d(v,u)}$$

if u is not an isolated point of X: $|\nabla f|(u) = 0$ if u is an isolated point of X.

If $u \in D(f)$, we say that "u is critical from below for f" if $|\nabla f|(u) = 0$, namely if u is isolated or if

$$\liminf_{v \to u} \frac{f(v)-f(u)}{d(v,u)} \geq 0.$$

(1.2) Remark

(a) If X is a normed linear space and f is differentiable at u, then we have evidently

$$|\nabla f|(u) = \|df(u)\|$$

(b) If, for instance, X is a Hilbert space, f is convex and lower semicontinuous and $u \in D(f)$, then

$$|\nabla f|(u) < +\infty \iff \partial f(u) \neq \emptyset$$

and we have

$$|\nabla f|(u) = \|\partial_0 f(u)\|$$

(for the definition of $\partial f(u)$ and $\partial_0 f(u)$ see [5]).

The following definition extends the classical one for solution of the equation

$$U' = -\,(\text{grad } f)(U).$$

(1.3) Definition

Let I be an interval with $\overset{\circ}{I} \neq \emptyset$.

A curve $U : I \to X$ is said to be of "(mean) maximal slope for f" if

(a) U is continuous

(b) $f \circ U(t) < +\infty$ whenever $t > \inf I$

(c) $d(U(t_2),U(t_1)) \leq \int_{t_1}^{*\,t_2} |\nabla f|(U(t))dt$, if $t_1 < t_2$ in I

(d) $f \circ U(t_2) - f \circ U(t_1) \leq -\int_{t_1}^{*\,t_2} (|\nabla f|(U(t)))^2 dt$, if $t_1 < t_2$ in I.

We remark that, if such a U exists, the function $|\nabla f|(U(t))$ is measurable.

Moreover the notions we have introduced are invariant under isometries of X.

Suppose now that f is defined on an open subset W of a Hilbert space H (or a Banach space with suitable assumptions: see [12]).

(1.4) Definition

Let u be in D(f). We designate as the "subdifferential of f at u" the (possibly empty) set $\partial^- f(u)$ of the α's of H such that

$$\liminf_{v \to u} \frac{f(v) - f(u) - (\alpha | v - u)}{\|v - u\|} \geq 0.$$

Since $\partial^- f(u)$ is convex and closed, if $\partial^- f(u) \neq \emptyset$ we can designate as the "subgradient of f at u" the element $\text{grad}^- f(u)$ of minimal norm in $\partial^- f(u)$.

It should be noted that u is critical from below for f if and only if $0 \in \partial^- f(u)$, namely $\text{grad}^- f(u) = 0$.

If f is differentiable or convex, these notions agree with the usual ones.

The following remark shows that the definitions we have given fit also the study of f on a constraint E.

If E is a subset of X, we denote by I_E the function defined by

$$I_E(u) = \begin{cases} 0 & \text{if } u \in E \\ +\infty & \text{if } u \in X \setminus E. \end{cases}$$

(1.5) Remark

Consider another function $g: W \to \mathbb{R} \cup \{+\infty\}$. Then

(a) for every u in $D(f) \cap D(g)$, we have

$$\partial^-(f+g)(u) \supset \partial^- f(u) + \partial^- g(u)$$

(b) if g is real and differentiable at u, we have

$$\partial^-(f+g)(u) = \partial^- f(u) + \text{grad } g(u)$$

(c) in particular, if E is a subset of H and g is real and differentiable at $u \in W \cap E$, we have

$$\partial^-(I_E + g)(u) = \partial^- I_E(u) + \text{grad } g(u)$$

and, if P is the orthogonal projection on the convex closed cone $\partial^- I_E(u)$, then

$$\text{grad}^-(I_E + g)(u) = \text{grad } g(u) + P(-\text{grad } g(u)).$$

We notice that, if E is, for instance, a submanifold of H without boundary, then $\partial^- I_E(u)$ is the set of vectors which are orthogonal to the tangent subspace to E at u.

(1.6) Remark

Let Ω be an open subset of $\mathbb{R}^n$ and let

$$\mathcal{E}(u) = \frac{1}{2} \int_\Omega |Du|^2 dx \text{ if } u \in H_0^1(\Omega).$$

(a) First of all it is obvious that we have

$$\partial^- \mathcal{E}(u) = \{u\}$$

with respect to the scalar product $\int_\Omega Du \cdot Dv\, dx$, as E is differentiable and grad $E(u) = u$ (in this case the equation $U' = -\text{grad}\, E(U)$ is trivial).

(b) Now we look at the problem in $L^2(\Omega)$, setting

$$\bar{E}(u) = \begin{cases} E(u) & \text{if } u \in H_0^1(\Omega) \\ +\infty & \text{if } u \in L^2(\Omega) \setminus H_0^1(\Omega) \end{cases}$$

As is well known, with respect to the usual scalar product of $L^2(\Omega)$ we have, if $u \in H_0^1(\Omega)$,

$\partial^- \bar{E}(u) \neq \emptyset \iff \Delta u \in L^2(\Omega)$ in the distribution sense and

$\partial^- \bar{E}(u) = \{-\Delta u\}$.

(c) Suppose now that Ω is bounded and regular. Let ϕ be a function in $H^2(\Omega)$ with $\phi \leqq 0$ on $\partial\Omega$.
Consider $\bar{E} + I_K : L^2(\Omega) \to \mathbb{R} \cup \{+\infty\}$, where

$$K = \{u \in L^2(\Omega) \mid u \geqq \phi \quad \text{a.e.}\}.$$

With respect to the scalar product of $L^2(\Omega)$ we have

$$\partial^-(\bar{E} + I_K)(u) \neq \emptyset \iff u \in H^2(\Omega), \text{ for } u \in H_0^1 \cap K$$

and moreover

$$\partial^-(\bar{E} + I_K)(u) = \partial^- \bar{E}(u) + \partial^- I_K(u).$$

Essentially these facts can be reduced to the following one:

$$\partial^-(\bar{E} + I_K)(u) \neq \emptyset \Rightarrow u \in H^2, \text{ for } u \in H_0^1 \cap K$$

(see [6], [22]).
Finally, if $\alpha \in L^2$ and $u \in H^2 \cap H_0^1 \cap K$, we have

$$\alpha \in \partial^-(\bar{E} + I_K)(u) \iff \begin{cases} \Delta u + \alpha = 0 & \text{in } \{x \mid u(x) > \phi(x)\} \\ \Delta u + \alpha \leqq 0 & \text{in } \{x \mid u(x) = \phi(x)\} \end{cases}$$

(1.7) Remark

It is evident that

$$\alpha \in \partial^- f(u) \Rightarrow |\nabla f|(u) \leq \|\alpha\| .$$

(1.8) Proposition

Suppose that for every u in D(f), with $|\nabla f|(u) < +\infty$, we have $\partial^- f(u) \neq \emptyset$ and $\|\text{grad}^- f(u)\| = |\nabla f|(u)$ (this assumption is true for a large class of functions: see [11], [12]).

Then, if $U : I \to W$ is a curve of maximal slope for f, we have

$$U'(t) = -\text{grad}^- f(U(t)) \quad \text{a.e. on } I.$$

(1.9) Remark

If $U : I \to W$ is a curve satisfying the following:

(a) U is continuous on I and absolutely continuous on compact subsets of $\mathring{I}$

(b) $f \circ U$ is non-increasing

(c) $U'(t) \in -\partial^- f(U(t))$ a.e. on I,

then U is a curve of maximal slope for f.

Several other definitions and related properties were introduced and studied in [12], [27].

2. RECALLS ON EXISTENCE THEOREMS WITH COMPACTNESS ASSUMPTIONS

I discuss here some of the results that were announced in [12] and proved in [28].

Let (X,d) be a metric space and $f : X \to \mathbb{R} \cup \{+\infty\}$ a function.

(2.1) Definition

The function f is said to be "locally coercive" if, for every $\rho > 0$, $c \in \mathbb{R}$, $u \in D(f)$, the set $\{v \in X \mid d(u,v) \leq \rho, f(v) \leq c\}$ is compact.

We begin with a theorem concerning continuous functions.

(2.2) Theorem

Suppose that

(a) the restriction of f to D(f) is continuous;

(b) f is locally coercive;

(c) for every u in D(f), we have:

$$|\nabla f|(u) \leqq \liminf_{\substack{v \to u \\ f(v) \leqq f(u)}} |\nabla f|(v).$$

Then for every u_0 in D(f), there exist $T > 0$ and a curve $U : [0,T] \to X$ of maximal slope for f such that $U(0) = u_0$.

This theorem allows us to obtain a result concerning a meaningful class of lower semicontinuous functions, by considering the function $\mathcal{G}_f : X \times \mathbb{R} \to \mathbb{R} \cup \{+\infty\}$ defined by

$$\mathcal{G}_f(v,r) = \begin{cases} r, & \text{if } f(v) \leq r \\ +\infty, & \text{if } f(v) > r. \end{cases}$$

It should be noted that the restriction of $\mathcal{G}_f$ to its domain is (Lipschitz) continuous.

One can deduce the following statement.

(2.3) Theorem

Suppose that

(a) f is locally coercive;

(b) for every u in D(f), we have:

$$|\nabla f|(u) \leq \liminf_{\substack{v \ u \\ f(v) \leq (f(u)}} |\nabla f|(v)$$

(c) for every u in D(f) and for every sequence $(u_n)_n$ such that $\lim_n u_n = u$, $f(u) < \lim_n f(u_n) < +\infty$, we have

$$\lim_n |\nabla f|(u_n) = +\infty.$$

Then for every u_0 in $D(f)$, there exist $T > 0$ and a curve $(V,R):[0,T] \to X \times \mathbb{R}$ of maximal slope for G_f such that $V(0) = u_0$, $R(0) = f(u_0)$.

It should be noted that the curve (V,R) can be regarded as a curve of maximal slope for f in a more general sense than (2.1), as for every curve U of maximal slope for f, there exists a curve (V,R) of maximal slope for G_f such that V is obtained from U by a suitable change of variable.

In certain conditions the converse is true (see [27]).

Therefore, there arises the problem of characterization of the functions which verify the hypotheses of (2.3) and also allow the passage from (V,R) to U.

A first significant class is pointed out in the following theorem

(2.4) Theorem

Suppose that

(a) f is locally coercive

(b) there exists a function $b : X \times \mathbb{R} \to \mathbb{R}$, bounded on bounded sets, and a continuous function $\omega : X \times X \to \mathbb{R}$ with $\omega(u,u) = 0$ such that

$$f(v) \geqq f(u) - |\nabla f|(u)d(v,u) - b(u,f(u))\omega(v,u)d(v,u)$$

whenever $v,u \in D(f)$ and $|\nabla f|(u) < +\infty$.

Then for every u_0 in $D(f)$, there exist $T > 0$ and a curve $U : [0,T] \to X$ of maximal slope for f such that $U(0) = u_0$.

It should be noted that the hypotheses of the previous theorem are, for instance, satisfied by the function $f : \mathbb{R}^n \to \mathbb{R} \cup \{+\infty\}$ defined by

$$f = g + g_0 + I_{\bar{\Omega}}$$

where g and g_0 are two real functions on $\mathbb{R}^n$, g of class C^1, g_0 convex, and Ω is an open subset of $\mathbb{R}^n$ with $\partial\Omega$ of class C^1.

It should be noted that, if Ω is not convex, f is not the sum of a convex function and a smooth function, as the set $D(f)$ is not convex.

The hypotheses of the theorem are also verified by the functional treated in the next section, where the geodesics with respect to an obstacle are studied.

For this aim, it is useful to state the following theorem.

(2.5) Theorem

Let H be a Hilbert space and $f : H \to \mathbb{R} \cup \{+\infty\}$ a locally coercive function.

Suppose that there exist two continuous functions $b : H \times \mathbb{R} \to \mathbb{R}$, $\omega : H \times H \to \mathbb{R}$, with $\omega(u,u) = 0$, such that

$$f(v) \geq f(u) + (\text{grad}^- f(u) \mid v-u) - b(u,f(u))\omega(v,u) \, \|v-u\|$$

whenever $u,v \in D(f)$ and $\partial^- f(u) \neq \emptyset$.

Then for every u_0 in $D(f)$, there exist $T > 0$ and a curve $U:[0,T] \to H$ of maximal slope for f such that $U(0) = u_0$.

Moreover we have

$$U'(t) = -\text{grad}^- f(U(t)) \quad \text{a.e. on } [0,T[.$$

3. GEODESICS WITH RESPECT TO AN OBSTACLE: EVOLUTION EQUATION AND MULTIPLICITY

The problem of geodesics "with respect to an obstacle", namely geodesics on manifolds with a boundary, is the subject of some recent studies (see for instance [1], [4], [45]) where some local properties, among other things, are analysed.

In [24] we found it useful to treat this subject in the framework of Sections 1 and 2. This allowed the study of the evolution equation associated with such curves, giving a result of global type.

In fact, in [24] it is shown that the classical multiplicity theorem (see [32], [44]), concerning the geodesics joining two points on a manifold without boundary, is still true, even in the presence of an obstacle; this theorem represents one of the most interesting and brilliant results of Morse theory and Lusternik-Schnirelman theory.

We observe that this study can be developed also in the framework of the ϕ-convex functions treated in Section 4, as is shown in [19].

Other studies concerning geodesics on manifolds with boundary have been developed in [34] to [43]. In particular, in [36] an expected result of local uniqueness is proved.

I refer next to the main results of [24].

We consider, for the sake of simplicity, a smooth function $\psi : \mathbb{R}^n \to \mathbb{R}$ such that the set $D = \{x \in \mathbb{R}^n \mid \psi(x) \geq 0\}$ is compact and non-empty and such that grad $\psi(x) \neq 0$ whenever $\psi(x) = 0$. In this case, the manifold with boundary is $\mathbb{R}^n \setminus \mathring{D}$; a more general case is treated in [34] and [39].

(3.1) Definition

A curve $\gamma:[0,1] \to \mathbb{R}^n \setminus \mathring{D}$ is said to be a geodesic with respect to the obstacle D if

(a) γ and γ' are absolutely continuous;

(b) there exists $\lambda:[0,1] \to [0, +\infty[$ such that

$$\gamma''(s) = \begin{cases} 0 & \text{a.e. on } \{s|\psi(\gamma(s)) < 0\} \\ \lambda(s)\nu(\gamma(s)) & \text{a.e. on } \{s|\psi(\gamma(s)) = 0\} \end{cases}$$

where $\nu(x) = |\text{grad } \psi(x)|^{-1} \text{ grad } \psi(x)$.

It is easy to check that in general γ is not of class C^2.

Let A and B be two points in $\mathbb{R}^n \setminus \mathring{D}$.

The following theorem shows that the geodesics with respect to D, joining A and B, are the critical points from below for the functional $f:L^2(0,1;\mathbb{R}^n) \to \mathbb{R} \cup \{+\infty\}$ defined by

$$f(\gamma) = \begin{cases} \frac{1}{2}\int_0^1 |\gamma'|^2 ds, & \text{if } \gamma \in X \\ +\infty \quad , & \text{if } \gamma \in L^2(0,1;\mathbb{R}^n) \setminus X \end{cases}$$

where $X = \{\gamma \in H^{1,2}(0,1;\mathbb{R}^n) | \gamma(0) = A, \gamma(1) = B, \psi(\gamma(s)) \leqq 0 \text{ on } [0,1]\}$.

In X we shall consider the scalar product of L^2.

(3.2) Theorem

Let γ be an element of X. Then

(a) $|\nabla f|(\gamma) < +\infty \Leftrightarrow \gamma \in H^{2,2}(0,1;\mathbb{R}^n) \Leftrightarrow \partial^- f(\gamma) \neq \emptyset$;

(b) if $|\nabla f|(\gamma) < +\infty$, then $|\nabla f|(\gamma) = \|\text{grad}^- f(\gamma)\|$ and

$$\text{grad}^- f(\gamma) = \begin{cases} -\gamma'' & \text{a.e. on } \{s|\psi(\gamma(s)) < 0\} \\ -\{\gamma''-[\gamma''\cdot(\nu\circ\gamma)]\nu 0 \quad \nu\circ\gamma\}, & \text{a.e. on } \{s|\psi(\gamma(s) = 0\}; \end{cases}$$

(c) in particular γ is a geodesic with respect to D if and only if $|\nabla f|(\gamma) = 0$;

(d) there exists $K \in \mathbb{R}$ such that, if γ and $\gamma + \delta$ are in X and $\partial^- f(\gamma) \neq \emptyset$, then

$$f(\gamma+\delta) \geq f(\gamma) + \int_0^1 \operatorname{grad}^- f(\gamma) \cdot \delta \, ds - K(f(\gamma))^2 \int_0^1 |\delta|^2 \, ds;$$

(e) for every γ_0 in X, there exists a unique curve $U : [0, +\infty[\to X$ of maximal slope for f such that $U(0) = \gamma_0$; moreover, if $(\gamma_h)_h$ is a sequence in X which converges to $\gamma \in X$ in $H^{1,2}(0,1;\mathbb{R}^n)$-metric, then, denoting by U_h and U the curves of maximal slope with starting points γ_h, γ, we have that $(U_h)_h$ converges to U uniformly on compact subsets of $[0,+\infty[$ in $H^{1,2}(0,1:\mathbb{R}^n)$-metric.

In order to get a multiplicity theorem, it is necessary to combine the results we have just given with information on the Lusternik-Schnirelman category of X and a theorem which relates such a category to the number of critical points from below for f.

The following statement transfers the well known techniques of Lusternik and Schnirelman (see [23]) to this framework. A similar result concerning locally Lipschitz continuous functions can be found in [8].

Let (Y,d) be a metric space and $g:Y \to \mathbb{R} \cup \{+\infty\}$ a lower semicontinuous function.

We consider in D(g) the "graph metric" d^* associated with g in the following way:

$$d^*(u,v) = d(u,v) + |g(u) - g(v)|$$

and we denote by Y^* the metric space $(D(g),d^*)$.

(3.3) <u>Theorem</u>

Suppose that

(a) g is bounded from below;

(b) the metric space Y^* is locally contractible;

(c) (Palais-Smale hypothesis) for every sequence $(u_h)_h$ in D(g) with $\sup_h g(u_h) < +\infty$, $\lim_h |\nabla g|(u_h) = 0$, there exists a subsequence $(u_{h_k})_k$

converging in d^* to an element u in D(g) with $|\nabla g|(u) = 0$;

(d) for every u in D(g), there exists a unique curve $U(u,\cdot):[0, +\infty[\to D(g)$ of maximal slope for g (with respect to d) such that $U(u,0) = u$ and such that

$$U : Y^* \times [0, +\infty[\to Y^*$$

is continuous.

Then, if the Lusternik-Schnirelman category of Y^* is k (possibly $k = +\infty$), the function g has at least k critical points from below.

(3.4) Remark

The graph metric associated with f induces in X the topology of $H^{1,2}(0,1;\mathbb{R}^n)$.

(3.5) Theorem

If $n \geq 2$ and A, B are in the unbounded connected component of $\mathbb{R}^n \setminus \mathring{D}$, then the category of X in $H^{1,2}$ is equal to $+\infty$.

(3.6) Theorem

If $n \geq 2$ and A, B are in the unbounded connected component of $\mathbb{R}^n \setminus \mathring{D}$, then there exist infinitely many geodesics with respect to D, joining A and B, with arbitrarily large length.

4. RECALLS ON ϕ-CONVEX FUNCTIONS

I give in this section some elements of the theory developed in the variational case, without compactness hypotheses. The papers on (p,q)-convex functions ([13, [17], [18]) (where (p,q)-monotone operators for the non-variational case are also considered) are oriented in this direction.

Later, we pointed out and studied in [11] and [20] (and in [15] for a non autonomous case) the class of ϕ-convex functions which extends the previous one.

The results that follow can be found in [11] and [20], together with several others which analyse local properties and limit properties with respect to Γ-convergence.

We still denote by H a Hilbert space and by $f:H \to \mathbb{R} \cup \{+\infty\}$ a function.

(4.1) Definition

Let $\phi : D(f) \times \mathbb{R} \times \mathbb{R} \to \mathbb{R}$ be a continuous function. The function f is said to be ϕ-convex if

$$f(v) \geqq f(u) + (\alpha | v-u) - \phi(u, f(u), \|\alpha\|) \|v-u\|^2$$

whenever $u, v \in D(f)$, $\partial^- f(u) \neq \emptyset$, $\alpha \in \partial^- f(u)$.

If $\phi(u, f(u), \|\alpha\|) \leqq \phi_0(u, f(u))(1 + \|\alpha\|^r)$, where $r \geqq 0$ and ϕ_0 is continuous, the function f is said to be ϕ-convex of order r.

We remark that the function f is not required to be subdifferentiable at some point.

For the study of the properties of this class of functions and, in particular, for the existence of the curves of maximal slope, the following result is essential.

(4.2) Theorem (existence of the minimum point)

Suppose that f is ϕ-convex and lower semicontinuous. Let u be in D(f) with $\partial^- f(u) \neq \emptyset$ and let $B(u, \delta)$ be an open ball where f is bounded from below. Then the function

$$v \mapsto f(v) + \mu \|v-u\|^2$$

has a unique minimum point in $B(u, \delta)$ provided that μ is large enough.

A first consequence of this theorem is the following one

(4.3) Proposition

If $\phi \equiv 0$ and f is lower semicontinuous and ϕ-convex then f is convex.

(4.4) Theorem (existence of the curve of maximal slope)

(a) If f is lower semicontinuous and ϕ-convex, then for every u_0 in D(f) with $\partial^- f(u_0) \neq \emptyset$, there exist $T > 0$ and a unique Lipschitz continuous curve $U : [0, T[\to H$ of maximal slope for f such that $U(0) = u_0$ and we have $\forall t \in [0, T[$, $\partial^- f(U(t)) \neq \emptyset$ and $U'_+(t) = -\text{grad}^- f(U(t))$.

(b) If f is lower semicontinuous and ϕ-convex of order two then u_0 can be chosen in D(f) and a continuous dependence on initial data holds.

5. EIGENVALUES OF LAPLACE OPERATOR WITH RESPECT TO AN OBSTACLE: EVOLUTION EQUATION AND MULTIPLICITY

In this section, I mention an evolution theorem obtained by studying (see [9], [26]) the functional f_1, defined in the introduction, by means of the theory of ϕ-convex functions.

This result allowed us to prove (see again [9], [26]) the existence of infinitely many "eigenvalues with respect to an obstacle", using that extension of the Lusternik-Schnirelman techniques to non regular functions already employed in [24].

Results, associated with the first eigenvalue, obtained by the study of the minimum of f_1 are given in [2], [3], [30].

For the well known result in the nonlinear case without an obstacle, see [7].

We point out that, even if in (5.4) the nonlinearity g does not appear, the presence of the obstacle forces us to use nonlinear techniques in order to get theorems (5.9) and (5.10).

In the following we consider a bounded regular open subset Ω of $\mathbb{R}^n$, two functions ϕ_1, ϕ_2 defined on Ω, a function $G : \Omega_x \times \mathbb{R}_s \to \mathbb{R}$ and a number $\rho > 0$.

We shall consider the following (simplified) hypotheses:

(5.1) for a.e. $x \in \Omega$, $G(x,\cdot)$ is of class C^2, for every $s \in \mathbb{R}$, $G(\cdot,s)$ is measurable, $G(x,0) = 0$, $\frac{\partial G}{\partial s}(\cdot,0) \in L^2(\Omega)$, $\frac{\partial^2 G}{\partial s^2}$ is bounded on $\Omega \times \mathbb{R}$;

(5.2) ϕ_1 and ϕ_2 are in $H^2(\Omega) \cap C(\Omega)$ and $\phi_1 \leqq \phi_2$ on Ω, $\phi_1 \leqq 0 \leqq \phi_2$ on $\partial\Omega$;

(5.3) (a) the open sets $\{x \in \Omega | \phi_1(x) < 0\}$, $\{x \in \Omega | \phi_2(x) > 0\}$ are connected (for instance $\phi_1 < 0 < \phi_2$ in Ω if Ω is connected);

(b) if $K = \{u \in H_0^1(\Omega) | \phi_1 \leqq u \leqq \phi_2,$ a.e. in $\Omega\}$ and $S_\rho = \{u \in L^2(\Omega) | \int_\Omega u^2 dx = \rho^2\}$, then

$$\inf\{\int_\Omega u^2 dx | u \in K\} < \rho^2 < \sup\{\int_\Omega u^2 dx | u \in K\},$$

$$\rho^2 \neq \int_\Omega \phi_1^2 dx, \quad \rho^2 \neq \int_\Omega \phi_2^2 dx.$$

We set $g = \frac{\partial G}{\partial s}$ and we introduce the following notations:

(5.4) if $u \in H^2(\Omega) \cap K$ and $\lambda \in \mathbb{R}$, set

$$-A(\lambda,u) = \begin{cases} \lambda u + \Delta u - g(x,u) & \text{if } \phi_1(x) < u(x) < \phi_2(x) \\ [\lambda u + \Delta u - g(x,u)]^+ & \text{if } \phi_1(x) = u(x) < \phi_2(x) \\ -[\lambda u + \Delta u - g(x,u)]^- & \text{if } \phi_1(x) < u(x) = \phi_2(x) \\ 0 & \text{if } \phi_1(x) = u(x) = \phi_2(x) \end{cases}$$

(5.5) We denote by $h : L^2(\Omega) \to \mathbb{R} \cup \{+\infty\}$ the functional defined by

$$h(u) = \begin{cases} \frac{1}{2}\int_\Omega |Du|^2 dx + \int_\Omega G(x,u)dx, & \text{if } u \in H_0^1(\Omega) \\ +\infty & , \text{if } u \in L^2(\Omega)\setminus H_0^1(\Omega), \end{cases}$$

where $Du = (D_1u,\dots,D_nu)$.

Finally we set $f = h + I_K + I_{S_\rho} : L^2(\Omega) \to \mathbb{R} \cup \{+\infty\}$.

I recall, in this section, some of the results of [9], [26].

(5.6) Proposition

In the hypotheses (5.1'), (5.2), (5.3), let $u \in K \cap S_\rho$, $\alpha \in L^2(\Omega)$. Then

(a) $\partial^- f(u) \neq \emptyset \iff u \in H^2(\Omega) \iff |\nabla f|(u) < +\infty$;

(b) $\partial^- f(u) = \partial^- h(u) + \partial^- I_K(u) + \partial^- I_{S_\rho}(u)$;

(c) $\alpha \in \partial^- f(u) \iff \exists\, \lambda \in \mathbb{R}$ such that

$$\int_\Omega Du\, D(v-u)dx + \int_\Omega g(x,u)(v-u)dx - \int_\Omega \lambda u(v-u)dx \geq \int_\Omega \alpha(v-u)dx, \ \forall v \in K;$$

(d) if $u \in H^2(\Omega)$, then $\text{grad}^- f(u) = A(\lambda,u)$ for a suitable real λ.

In particular, $0 \in \partial^- f(u) \iff \exists\, \lambda \in \mathbb{R}: A(\lambda,u) = 0$, namely $\lambda u \in \partial^-(h + I_K)(u)$, that is

(5.7) $$\int_\Omega Du\, D(v-u)dx + \int_\Omega g(x,u)(v,u)dx \geq \int_\Omega \lambda u(v-u)dx, \quad \forall v \in K.$$

(5.8) Proposition

In the hypotheses (5.1), (5.2), (5.3) let I be an open interval and $U : I \to L^2(\Omega)$ an absolutely continuous curve, such that $U(t) \in K \cap S_\rho$ $\forall t \in I$. Then the following facts are equivalent:

(a) U is a curve of maximal slope for f;

(b) there exists a bounded function on compact sets $\lambda : I \to \mathbb{R}$ such that for almost every t in I

$$\int_\Omega U'(t)(v-U(t))dx + \int_\Omega DU(t)D(v-U(t))dx$$

$$+\int_\Omega g(x,U(t))(v-U(t))dx - \int_\Omega \lambda(t)U(t)(v-U(t))dx \geqq 0 \qquad \forall v \in K;$$

(c) for every t in I, $U(t) \in H^2(\Omega)$ and $U'_+(t)$ exists; moreover there exists a bounded function on compact sets $\lambda : I \to \mathbb{R}$ such that

$$U'_+(t) + A(\lambda(t), U(t)) = 0, \text{ on } I.$$

(5.9) Theorem

In the hypotheses (5.1), (5.2), (5.3), we have that for every u_0 in $K \cap S_\rho$, there exists a unique curve $U : [0, +\infty[\to L^2(\Omega)$ of maximal slope for f such that $U(0) = u_0$.

In the proof of the previous statements an essential role is played by the fact that the "constraint" $K \cap S_\rho$ has nice properties (see [9]), since K and S_ρ are not "mutually tangential", in a suitable sense, by virtue of hypothesis (5.3).

In particular, theorem (5.9) follows from the fact that the assumptions imply that f is ϕ-convex of order 1.

The following multiplicity result is based on the previous theorem and theorem (3.3).

(5.10) Theorem

In the hypotheses (5.1), (5.2), (5.3), suppose that

$$g(x,-s) = -g(x,s), \quad \phi_1 = -\phi_2.$$

Then there exist infinitely many pairs (λ,u) with $\lambda \in \mathbb{R}$ and $u \in H^2(\Omega) \cap K \cap S_\rho$ such that

$$A(\lambda,u) = 0.$$

The set of such λ's is bounded from below and unbounded from above.

REFERENCES

[1] R. Alexander, S. Alexander, "Geodesics in Riemannian manifolds-with-boundary", Indiana Univ. Math. J. 30 (1981), 481-488.

[2] V. Benci, "Positive solutions of some eigenvalue problems in the theory of variational inequalities", J. Math. Anal. Appl. 61 (1977), 165-187.

[3] V. Benci, A.M. Micheletti, "Su un problema di autovalori per disequazioni variazionali", Ann. Mat. Pura Appl. (4) 107 (1975), 359-371.

[4] I.D. Berg, "An estimate on the total curvature of a geodesic in Euclidean 3-space-with-boundary", Geom. Dedicata, 13 (1982), 1-6.

[5] H. Brezis, "Opérateurs maximaux monotones et semigroupes de contractions dans les espaces de Hilbert", North-Holland Mathematics Studies, N. 5, Notas de Matemàtica (50), Amsterdam-London, 1973.

[6] H. Brezis, G. Stampacchia, "Sur la régularité dela solution d'inéquations elliptiques", Bull. Soc. Math. France 96 (1968), 153-180.

[7] F.E. Browder, "Infinite dimensional manifolds and non-linear elliptic eigenvalue problems", Ann. of Math. (2) 82 (1965), 459-477.

[8] K.C. Chang, "Variational methods for non-differentiable functionals and their applications to partial differential equations", J. Math. Anal. Appl. 80 (1981), 102-129.

[9] G. Chobanov, A. Marino, D. Scolozzi, to appear.

[10] F.H. Clarke, "Optimization and non-smooth analysis", John Wiley, New York, 1983.

[11] E. De Giorgi, M. Degiovanni, A. Marino, M. Tosques, "Evolution equations for a class of non-linear operators", Atti Accad. Naz. Lincei Rend. Cl. Sci. Fis. Mat. Natur. (8) 75 (1983), 1-8.

[12] E. De Giorgi, A. Marino, M. Tosques, "Problemi di evoluzione in spazi metrici e curve di massima pendenza", Atti Accad. Naz. Lincei Rend. Cl. Sci. Fis. Mat. Natur. (8) 68 (1980), 180-187.

[13] E. De Giorgi, A. Marino, M. Tosques, "Funzioni (p,q)-convesse", Atti Accad. Naz. Lincei Rend. Cl. Sci. Fis. Mat. Natur. (8) 73 (1982), 6-14.

[14] M. Degiovanni, "Alcuni sistemi iperbolici non lineari del primo ordine visti come problemi di evoluzione in presenza di più norme", Boll. Un. Mat. Ital. B (6) 1 (1982), 47-73.

[15] M. Degiovanni, "Parabolic equations with nonlinear time-dependent boundary conditions", Ann. Mat. Pura Appl. (4) 141 (1985), 223-264.

[16] M. Degiovanni, A. Marino, "Alcuni problemi di autovalori e di biforcazione rispetto ad un ostacolo", Equazioni differenziali e calcolo delle variazioni (Pisa, 1985), pp. 71-93, ETS, Pisa, 1985.

[17] M. Degiovanni, A. Marino, M. Tosques, "General properties of (p,q)-convex functions and (p,q)-monotone operators", Ricerche Mat. 32 (1983), 285-319.

[18] M. Degiovanni, A. Marino, M. Tosques, "Evolution equations associated with (p,q)-convex functions and (p,q)-monotone operators", Ricerche Mat. 33 (1984), 81-112.

[19] M. Degiovanni, A. Marino, M. Tosques, "Critical points and evolution equations", Multifunctions and Integrands (Catania, 1983), pp. 184-192, Lecture Notes in Math. 1091, Springer, Berlin-New York, 1984.

[20] M. Degiovanni, A. Marino, M. Tosques, "Evolution equations with lack of convexity", Nonlinear Anal. 9 (1985), 1401-1443.

[21] M. Degiovanni, M. Tosques, "Evolution equations for (ϕ,f)-monotone operators", Boll. Un. Mat. Ital. B, in press.

[22] H. Lewy, G. Stampacchia, "On the regularity of the solution of a variational inequality", Comm. Pure Appl. Math. 22 (1969), 153-188.

[23] L.A. Lusternik, L. Schnirelman, "Méthodes topologiques dans les problèmes variationels", Gauthier - Villars, Paris, 1934.

[24] A. Marino, D. Scolozzi, "Geodetiche con ostacolo", Boll. Un. Mat. Ital. B (6) 2 (1983), 1-31.

[25] A. Marino, D. Scolozzi, "Punti inferiormente stazionari ed equazioni di evoluzione con vincoli unilaterali non convessi", Rend. Sem. Mat. Fis. Milano 52 (1982), 393-414.

[26] A. Marino, D. Scolozzi, "Autovalori dell'operatore di Laplace ed equazioni di evoluzione in presenza di ostacolo", Problemi differenziali e teoria dei punti critici (Bari, 1984), pp. 137-155, Pitagora, Bologna, 1984.

[27] A. Marino, M. Tosques, "Curves of maximal slope for a certain class of non regular functions", Boll. Un. Mat. Ital. B(6) 1 (1982), 143-170.
[28] A. Marino, M. Tosques, "Existence and properties of the curves of maximal slope", Istituto di Matematica, Università di Pisa, preprint 1981.
[29] E. Mitidieri, M. Tosques, "Volterra integral equations associated to a class of nonlinear operators in Hilbert spaces", Preprint n. 121, Dipartimento di Matematica, Università di Pisa, 1985.
[30] R.C. Riddell, "Eigenvalue problems for nonlinear elliptic variational inequalities", Nonlinear Anal. 3 (1979), 1-33.
[31] R.T. Rockafellar, "Generalized directional derivatives and subgradients of nonconvex functions", Canad. J. Math. 32 (1980), 257-280.
[32] J.T. Schwartz, "Generalizing the Lusternik-Schnirelman theory of critical points", Comm. Pure Appl. Math. 17 (1964), 307-315.
[33] J.T. Schwartz, "Nonlinear functional analysis", Gordon & Breach, New York, 1969.
[34] D. Scolozzi, "Geodetiche su sottovarietà con bordo di $\mathbb{R}^n$", Istituto di Matematica, Università di Pisa, preprint, 1981.
[35] D. Scolozzi, "Una osservazione sulle geodetiche con ostacolo", Istituto di Matematica, Università di Pisa, preprint, 1981.
[36] D. Scolozzi, "Un risultato di locale unicità per le geodetiche su varietà con bordo", Boll. Un. Mat. Ital. B, in press.
[37] D. Scolozzi, "Un teorema di esistenza di una geodetica chiusa su varietà con bordo", Preprint n. 22, Dipartimento di Matematica, Università di Pisa, 1983.
[38] D. Scolozzi, "Convergenza di geodetiche con ostacolo", Preprint n. 50, Dipartimento di Matematica, Università di Pisa, 1984.
[39] D. Scolozzi, "Esistenza e molteplicità di geodetiche con ostacolo e con estremi variabili", Ricerche Mat. 33 (1984), 171-201.
[40] D. Scolozzi, "Molteplicità di curve con ostacolo e stazionarie per una classe di funzionali non regolari", Preprint n. 69, Dipartimento di Matematica, Università di Pisa, 1984.
[41] D. Scolozzi, "Esistenza di una curva chiusa stazionaria con ostacolo", Preprint n. 70, Dipartimento di Matematica, Università di Pisa, 1984.
[42] D. Scolozzi, "Un teorema di esistenza di una geodetica chiusa su varietà con bordo", Boll. Un. Mat. Ital. A, in press.

[43] D. Scolozzi, "Γ-convergenza di funzionali del tipo delle geodetiche", Preprint n. 93, Dipartimento di Matematica, Università di Pisa, 1984.

[44] J.P. Serre, "Homologie singulière des espaces fibrés", Ann. of Math. 54 (1951), 425-505.

[45] F.E. Wolter, "Interior metrics, shortest paths and loops in Riemannian manifolds with not necessarily smooth boundary", preprint.

A. Marino
Dipartimento di Matematica
Università di Pisa
56100 Pisa
Italy

F MURAT

A survey on compensated compactness

This survey presents a few results of compensated compactness. Since these results are concerned with weak lower semicontinuity properties of functionals, I believe that they are a suitable topic in a conference dedicated to the memory of Leonida Tonelli. The work presented here is the fruit of a long and friendly collaboration with Luc Tartar, although we have not yet written any joint paper on the topic. Therefore, I refer to our (individual) papers [1] to [7] for published results. The only new results presented here (Theorems 4 and 7.6) are extensions of previous results.

1. SETTING THE PROBLEM

Let Ω be an open set of $\mathbb{R}^N$; since this survey is only concerned with "local" properties, the boundedness and the smoothness of Ω are irrelevant. We consider functions u^ε defined in Ω with values in $\mathbb{R}^m$ $(m \geq 1)$; here ε is a parameter which takes its values in a sequence converging to zero (say $\varepsilon = 1/n$, $n \in \mathbb{N}$). Let us assume that

$$u^\varepsilon \longrightarrow u \text{ in } (L^\infty(\Omega))^m \text{ weak } *, \tag{1.1}$$

$$\begin{cases} \sum_{i=1}^{m} \sum_{j=1}^{N} a_{ijk} \dfrac{\partial u_i^\varepsilon}{\partial x_j} \text{ bounded in } L^2(\Omega) \\ \\ \text{(independently of } \varepsilon\text{) for } k = 1,\dots,n, \end{cases} \tag{1.2}$$

where the $a_{ijk} \in \mathbb{R}$ are given coefficients. We also consider a function f

$$f \in C^0(\mathbb{R}^m;\mathbb{R}) \tag{1.3}$$

Since f is continuous and u^ε bounded in $(L^\infty(\Omega))^m$, the functions $f(u^\varepsilon)$ are bounded in $L^\infty(\Omega)$. We will investigate the following problem:

<u>Question 1</u>: for a given set of coefficients a_{ijk}, which are the functions $f \in C^0(\mathbb{R}^m;\mathbb{R})$ such that:

$$f(u^\varepsilon) \longrightarrow f(u) \text{ in } L^\infty(\Omega) \text{ weak } * \tag{1.4}$$

for all sequences u^ε satisfying (1.1), (1.2)? □

In other terms we seek (all the) functions f associated with sequentially weakly continuous mappings from the space

$$Z_\infty = \{u | u \in (L^\infty(\Omega))^m, \sum_i \sum_j a_{ijk} \frac{\partial u_i}{\partial x_j} \in L^2(\Omega), 1 \leq k \leq n\} \tag{1.5}$$

into $L^\infty(\Omega)$; "weakly" means here that both Z_∞ and $L^\infty(\Omega)$ are endowed with their (natural) weak * topologies.

A natural extension of the previous problem is the question of the sequentially weak lower semicontinuity of the functionals associated with f (and considered as mappings from Z_∞ into $\mathcal{D}'(\Omega)$) in the following sense:

Question 2: For a given set of coefficients a_{ijk}, which are the functions $f \in C^0(\mathbb{R}^m;\mathbb{R})$ such that:

$$\begin{cases} \forall \phi \in \mathcal{D}(\Omega), \quad \phi \geq 0, \\ \liminf_{\varepsilon\to 0} \int \phi(x) f(u^\varepsilon(x))dx \geq \int_\Omega \phi(x)\, f(u(x))dx \end{cases} \tag{1.6}$$

for all sequences u^ε satisfying (1.1), (1.2)? □

Remark 1.1 The choice of the spaces $L^\infty(\Omega)$ in (1.1) and $L^2(\Omega)$ in (1.2) is slightly arbitrary. Since we consider all the functions $f \in C^0(\mathbb{R}^m;\mathbb{R})$ without any kind of growth assumption, $L^\infty(\Omega)$ is actually a natural choice in (1.1). When a growth condition is imposed on f, (1.1) will be replaced by the convenient L^p; for example we will replace $(L^\infty(\Omega))^m$ by $(L^2(\Omega))^m$ in (1.1) whenever f has quadratic growth. □

Let us immediately illustrate the previous questions by means of two "extreme" well-known examples (more interesting examples will be given in Section 7).

Example 1.1

$$a_{ijk} = 0 \quad \forall i, \forall j, \forall k \tag{1.7}$$

In this case (1.2) does not provide any information about the derivatives of the u^ε's. It is well known that $u \to f(u)$ is sequentially weakly continuous if and only if f is affine. Similarly, $u \to f(u)$ is weakly lower semicontinuous (in the sense of (1.6)) if and only if f is convex. □

Example 1.2 Let us choose the coefficients a_{ijk} in such a manner that all derivatives $\frac{\partial u_i^\varepsilon}{\partial x_j}$ $(1 \leq i \leq m,\ 1 \leq j \leq N)$ of u^ε are bounded in $L^2(\Omega)$. In other terms, the space Z_∞ defined by (1.5) reduces to

$$Z_\infty \equiv (L^\infty(\Omega))^m \cap (H^1(\Omega))^m. \tag{1.8}$$

Rellich's compactness theorem then implies that the mapping $u \to f(u)$ is sequentially weakly continuous for *all* continuous functions f. □

Our main goal here is to consider the intermediate cases between the first example (no information on the derivatives, f convex) and the second one (information on all derivatives, no restriction on f). When information is available on some fixed derivatives, which are the functions f associated to sequentially weakly continuous or weakly lower semicontinuous mappings?

To answer this question (at least partially) we need a new notation. For a given set of coefficients a_{ijk} we define:

$$\begin{cases} V = \{(\lambda,\xi) \mid \lambda \in \mathbb{R}^m,\ \xi \in \mathbb{R}^N - \{0\}, \\ \qquad \sum_i \sum_j a_{ijk}\, \lambda_i\, \xi_j = 0,\ k = 1,\ldots,n\}. \end{cases} \tag{1.9}$$

$$\begin{cases} \Lambda = \{\lambda \mid \lambda \in \mathbb{R}^m,\ \exists \xi \in \mathbb{R}^N - \{0\}, \\ \qquad \sum_i \sum_j a_{ijk}\, \lambda_i\, \xi_j = 0,\ k = 1,\ldots,n\}. \end{cases} \tag{1.10}$$

Remark 1.2 The cone Λ is merely the projection on $\mathbb{R}^m$ of the manifold V. A good understanding of V (and simplifications of a few computations) can be achieved by noting that if u satisfies:

$$\sum_i \sum_j a_{ijk} \frac{\partial u_i}{\partial x_j} = 0,\ k = 1,\ldots,n,$$

then its Fourier transform satisfies

$$(\mathrm{Re}\ \hat{u}(\xi),\xi) \in V, \ (\mathrm{Im}\ \hat{u}(\xi),\xi) \in V, \quad \forall \xi \in \mathbb{R}^N,$$

and conversely. □

In the case of the two previous examples we have:

Example 1.1 (case without derivatives)

$$Z_\infty = (L^\infty(\Omega))^m \text{ and } \Lambda = \mathbb{R}^m.$$

Example 1.2 (case with all derivatives)

$$Z_\infty = (L^\infty(\Omega))^m \cap (H^1(\Omega))^m \text{ and } \Lambda = \{0\}.$$

2. A FIRST NECESSARY CONDITION FOR WEAK LOWER SEMICONTINUITY

In this section a first necessary condition on f to answer questions 1 or 2 is given. Let us recall that the a_{ijk} are fixed, and that Λ is defined by (1.10).

Theorem 2.1 If the functional $u \to f(u)$ is sequentially weakly lower semi-continuous in the sense of (1.6), then f is convex on all segments [a,b] with $a-b \in \Lambda$, i.e.:

$$\begin{cases} \forall a,\ b \in \mathbb{R}^m \text{ with } b-a \in \Lambda, \\ t \in \mathbb{R} \to f(a + t(b-a)) \in \mathbb{R} \text{ is convex. } \square \end{cases} \tag{2.1}$$

Remark 2.1 If $f \in C^2(\mathbb{R}^m;\mathbb{R})$ the convexity of f on all segments [a,b] with $a-b \in \Lambda$ is equivalent to

$$f''(a)\lambda\lambda \geq 0, \ \forall a \in \mathbb{R}^m, \ \forall \lambda \in \Lambda . \quad \square \tag{2.2}$$

Theorem 2.2 If the mapping $u \to f(u)$ is sequentially weakly continuous from Z_∞ defined by (1.5) into $L^\infty(\Omega)$, then f is affine in the directions of Λ, i.e.:

$$\begin{cases} \forall a \in \mathbb{R}^m, \quad \forall \lambda \in \Lambda, \\ t \in \mathbb{R} \to f(a + t\lambda) \in \mathbb{R} \text{ is affine. } \square \end{cases} \tag{2.3}$$

Proof of Theorems 2.1 and 2.2

Let us consider $a,\ b \in \mathbb{R}^m$, $\xi \in \mathbb{R}^N$, $\xi \neq 0$, $t \in \mathbb{R}$, $0 < t < 1$ and a function

$\chi:\mathbb{R} \to \mathbb{R}$, periodic with period 1 defined by:

$$\chi(s) = \begin{cases} 1 & \text{if } 0 \leq s < t, \\ 0 & \text{if } t \leq s < 1. \end{cases}$$

Let u^ε be defined by:

$$u^\varepsilon(x) = \chi(\frac{x \cdot \xi}{\varepsilon})a + (1 - \chi(\frac{x \cdot \xi}{\varepsilon}))b, \quad \forall x \in \Omega \ . \tag{2.4}$$

This sequence is bounded in $(L^\infty(\Omega))^m$ and

$$u^\varepsilon \longrightarrow t\ a + (1-t)b \text{ in } (L^\infty(\Omega))^m \text{ weak } * \ . \tag{2.5}$$

Thus (1.1) is satisfied. But

$$\frac{\partial u_i}{\partial x_j} = (a-b)_i\ \xi_j\ \chi'(\frac{x \cdot \xi}{\varepsilon})$$

where $\chi'(s)$ is the periodic distribution on $\mathbb{R}$ with period 1 defined on the period $0 \leq s < 1$ by

$$\chi'(s) = \delta_{s=0} - \delta_{s=t}, \ (\delta = \text{Dirac mass}) \ .$$

The sequence u^ε defined by (2.4) satisfies (1.2) if (and only if)

$$\sum_i \sum_j a_{ijk}\ (a-b)_i\ \xi_j = 0, \quad k = 1,\dots,n,$$

i.e. if (and only if)

$$((a-b),\xi) \in V \ . \tag{2.6}$$

If $(a-b) \in \Lambda$, there exists ξ such that (2.6) holds true and the sequence u^ε defined by (2.4) satisfies (1.1) and (1.2). On the other hand

$$f(u^\varepsilon(x)) = \chi(\frac{x \cdot \xi}{\varepsilon})\ f(a) + (1 - \ \chi(\frac{x \cdot \xi}{\varepsilon}))f(b),$$

and thus:

$$f(u^\varepsilon) \longrightarrow t\, f(a) + (1-t)\, f(b) \text{ in } L^\infty(\Omega) \text{ weak } \star. \tag{2.7}$$

If the functional $u \to f(u)$ is weakly lower semicontinuous, (2.5) and (2.7) imply that:

$$f(ta + (1-t)b) \leq t\, f(a) + (1-t)\, f(b),$$

which gives (2.1). Similarly, (2.3) follows from the weak continuity of $u \to f(u)$. □

3. THIS NECESSARY CONDITION IS SUFFICIENT WHENEVER f IS QUADRATIC

In this section we prove that if f is a quadratic form, the necessary condition (2.1) of Theorem 2.1 is sufficient for the weak lower semicontinuity of $u \to f(u)$. Similarly, the necessary condition (2.3) of Theorem 2.2 is sufficient for the weak continuity of $u \to f(u)$. Thus, in the case of quadratic forms, questions 1 and 2 are completely solved. The answer is a simple algebraic condition.

Theorem 3.1 Let f be a quadratic form on $\mathbb{R}^m$ satisfying

$$f(\lambda) \geq 0, \quad \forall\lambda \in \Lambda \tag{3.1}$$

where Λ is defined by (1.10). Then the functional $u \to f(u)$ is weakly lower semicontinuous in the sense of (1.6). □

Remark 3.1 Let f be a quadratic form

$$\frac{1}{2}\, f''(a)\, \lambda, \lambda \; = f(\lambda), \; \forall a \in \mathbb{R}^m, \; \forall\lambda \in \mathbb{R}^m. \tag{3.2}$$

Then (see Theorem 2.1 and Remark 2.1) *in the case of quadratic forms condition (3.1) is necessary and sufficient for the weak lower semicontinuity* of $u \to f(u)$. Thus question 2 is answered with a very simple algebraic condition. Note that in this case the convexity of f reads as

$$f(\lambda) \geq 0, \forall\lambda \in \mathbb{R}^m \tag{3.3}$$

which greatly differs from (3.1) when the cone Λ is small. □

Actually we will prove a slightly stronger version of Theorem 3.1. Since f is quadratic, we allow the sequence u^ε to be bounded in $(L^2(\Omega))^m$ (in place of $(L^\infty(\Omega))^m$ in (1.1)); $f(u^\varepsilon)$ is then bounded in $L^1(\Omega)$. Moreover, we allow the derivatives $\sum_i \sum_j a_{ijk} \frac{\partial u_i^\varepsilon}{\partial x_j}$ to lie in a compact set (for the strong topology) of $H^{-1}(\Omega)$ (in place of a bounded set of $L^2(\Omega)$ in (1.2)); note that this information does not follow from the boundedness of u^ε in $(L^2(\Omega))^m$. Precisely we prove:

<u>Theorem 3.2</u> Let f be a quadratic form on $\mathbb{R}^m$ satisfying (3.1). For all sequences u^ε such that:

$$u^\varepsilon \longrightarrow u \text{ in } (L^2(\Omega))^m \text{ weak,} \tag{3.4}$$

$$\begin{cases} \sum_i \sum_j a_{ijk} \frac{\partial u_i^\varepsilon}{\partial x_j} \text{ belongs to a compact set of } H^{-1}(\Omega) \\ \text{(independently of } \varepsilon) \text{ for } k = 1,\ldots,n, \end{cases} \tag{3.5}$$

one obtains:

$$\begin{cases} \forall\phi \in \mathcal{D}(\Omega), \\ \liminf_{\varepsilon\to 0} \int \phi^2(x)\, f(u^\varepsilon(x))dx \geq \int_\Omega \phi^2(x)\, f(u(x))dx. \quad \square \end{cases} \tag{3.6}$$

As far as weak continuity is concerned, we prove

<u>Theorem 3.3</u> Let f be a quadratic form satisfying

$$f(\lambda) = 0, \quad \forall\lambda \in \Lambda. \tag{3.7}$$

Then the application $u \to f(u)$ is sequentially weakly continuous in the sense of (1.4). □

<u>Remark 3.2</u> Using the same argument as in Remark 3.1, and a slightly stronger version (similar to Theorem 3.2) of Theorem 3.3, we actually claim that *in the case of quadratic forms, condition (3.7) is equivalent to the sequential weak continuity* of the mapping $u \to f(u)$ for sequences u^ε satisfying (3.4) and (3.5), the mapping taking its values in $L^1(\Omega)$ endowed with the vague topology of measures. □

Proof of Theorems 3.1, 3.2 and 3.3

Considering f and -f, Theorem 3.3 reduces to a corollary of Theorem 3.1, which itself follows from Theorem 3.2. So let us prove Theorem 3.2.

Step 1 : Localization and Fourier transform

For any $\phi \in \mathcal{D}(\Omega)$, the sequence v^ε and the function v defined by:

$$v^\varepsilon = \begin{cases} \phi u & \text{in } \Omega \\ 0 & \text{in } \mathbb{R}^N \setminus \Omega \end{cases} \qquad v = \begin{cases} \phi u & \text{in } \Omega \\ 0 & \text{in } \mathbb{R}^N \setminus \Omega \end{cases}$$

satisfy:

$$\begin{cases} v^\varepsilon \dashrightarrow v \text{ in } (L^2(\mathbb{R}^N))^m \text{ weak}, & (3.8) \\ \sum_i \sum_j a_{ijk} \dfrac{\partial v_i^\varepsilon}{\partial x_j} \text{ compact in } H^{-1}(\mathbb{R}^N) & (3.9) \end{cases}$$

(independently of ε) $k = 1,\ldots,n$,

$$\operatorname{supp} v^\varepsilon \subset K, \tag{3.10}$$

where $K = \operatorname{supp} \phi$ is a compact set of $\mathbb{R}^N$.

Moreover because ϕ is quadratic,

$$\phi^2(x)\, f(u^\varepsilon(x)) = f(v^\varepsilon(x)),\ \phi^2(x)\, f(u(x)) = f(v(x))$$

and (3.6) reads as

$$\liminf_{\varepsilon\to 0} \int_{\mathbb{R}^N} f(v^\varepsilon(x))dx \geq \int_{\mathbb{R}^N} f(v(x))dx. \tag{3.11}$$

The problem is now reduced to the case $\Omega = \mathbb{R}^N$, together with a sequence v^ε with compact support and $\phi = 1$.

Let $\hat{v}^\varepsilon$ and $\hat{v}$ denote the Fourier transforms of v^ε and v and $\tilde{f}$ the Hermitian extension to $\mathbb{C}^m$

$$\tilde{f}(\lambda) = \sum_{j,j'=1}^{m} f_{jj'}\, \lambda_j \bar{\lambda}_{j'} \qquad (\lambda \in \mathbb{C}^m)$$

of the original quadratic form $f(\lambda) = \sum_{j,j'=1}^{m} f_{ij'}\, \lambda_j \lambda_{j'}$ (defined for $\lambda \in \mathbb{R}^m$).

Since we can always assume $f_{jj'} = f_{j'j}$ we have

$$\tilde{f}(\lambda) \in \mathbb{R}, \ \forall \lambda \in \mathbb{C}^m.$$

Application of the Plancherel-Parseval theorem and decomposition of the resulting integral in two parts yields, for any fixed real number $R > 0$:

$$\begin{cases} \int_{\mathbb{R}^N} f(v^\varepsilon(x))dx = \int_{\mathbb{R}^N} \tilde{f}(\hat{v}^\varepsilon(\xi))d\xi \\ = \int_{|\xi|\leq R} \tilde{f}(\hat{v}^\varepsilon(\xi))d\xi + \int_{|\xi|>R} \tilde{f}(\hat{v}^\varepsilon(\xi))d\xi. \end{cases} \tag{3.12}$$

Step 2 : study of the term at finite distance

By virtue of (3.10)

$$\hat{v}^\varepsilon(\xi) = \int_{\mathbb{R}^N} v^\varepsilon(x)\ e^{2i\pi x\cdot\xi}\ dx = \int_K v^\varepsilon(x)e^{2i\pi x\cdot\xi}\ dx$$

and (3.8) implies

$$\begin{cases} |\hat{v}^\varepsilon(\xi)| \leq \|v^\varepsilon\|_{L^1(K)} \leq M \\ \hat{v}^\varepsilon(\xi) \to \hat{v}(\xi) \end{cases} \quad \forall \xi \in \mathbb{R}^N,$$

where M is a constant independent of ε and ξ.

Thus for any fixed $R > 0$, Lebesgue's dominated convergence theorem yields:

$$\int_{|\xi|\leq R} \tilde{f}(\hat{v}^\varepsilon(\xi))d\xi \to \int_{|\xi|\leq R} \tilde{f}(\hat{v}(\xi))d\xi \quad \text{as } \varepsilon \to 0\ . \tag{3.13}$$

But

$$\begin{cases} \int_{|\xi|\leq R} \tilde{f}(\hat{v}(\xi))d\xi = \int_{\mathbb{R}^N} \tilde{f}(\hat{v}(\xi))d\xi - \int_{|\xi|>R} \tilde{f}(\hat{v}(\xi))d\xi \\ \qquad = \int_{\mathbb{R}^N} f(v(x))dx - \int_{|\xi|>R} \tilde{f}(\hat{v}(\xi))d\xi, \end{cases} \tag{3.14}$$

and since $\tilde{f}(\hat{v}) \in L^1(\mathbb{R}^N)$,

$$\int_{|\xi|>R} \tilde{f}(\hat{v}(\xi))d\xi \to 0 \text{ as } R \to +\infty. \tag{3.15}$$

Thus passing to the limit in (3.12) reduces to controlling the term at infinity $\int_{|\xi|>R} \tilde{f}(\hat{v}^{\varepsilon}(\xi))d\xi$.

<u>Step 3 : study of the term at infinity</u>

We begin with a lemma:

<u>Lemma 3.1</u> Let f be a quadratic form satisfying (3.1). Then for all $\delta > 0$ there exists $C_\delta > 0$ such that:

$$\begin{cases} \forall\xi \in \mathbb{R}^N - \{0\}, \quad \forall\lambda \in \mathbb{C}^m \\ \tilde{f}(\lambda) \geqq -\delta|\lambda|^2 - C_\delta \sum_k \Big|\sum_i \sum_j a_{ijk}\, \lambda_i \dfrac{\xi_j}{|\xi|}\Big|^2. \end{cases} \tag{3.16}$$ □

Let us prove this lemma by contradiction: if (3.16) does not hold true, then there exists $\delta_o > 0$ such that for all $\varepsilon > 0$:

$$\begin{cases} \exists\xi^\varepsilon \in \mathbb{R}^N, \ |\xi^\varepsilon| = 1, \quad \exists\lambda^\varepsilon \in \mathbb{C}^m, \ |\lambda^\varepsilon| = 1, \\ \tilde{f}(\lambda^\varepsilon) < -\delta_o\ 1 - \dfrac{1}{\varepsilon} \sum_k \Big|\sum_i \sum_j a_{ijk}\, \lambda_i\, \xi_j^\varepsilon\Big|^2. \end{cases}$$

Since $\tilde{f}(\lambda^\varepsilon)$ is bounded, a subsequence $(\lambda^{\varepsilon'},\xi^{\varepsilon'})$ of $(\lambda^\varepsilon,\xi^\varepsilon)$ satisfies:

$$\begin{cases} \lambda^{\varepsilon'} \to \lambda \quad \text{in } \mathbb{C}^m, \\ \xi^{\varepsilon'} \to \xi \quad \text{in } \mathbb{R}^N, \quad |\xi| = 1, \\ \sum_i \sum_j a_{ijk}\, \lambda_i\, \xi_j = 0, \ k = 1,\dots,n, \\ \tilde{f}(\lambda) \leqq -\delta_o. \end{cases}$$

But (3.1) implies that

$$\tilde{f}(\lambda) \geqq 0 \quad \forall\lambda \in \Lambda + i\Lambda$$

which gives a contradiction. Lemma 3.1 is proved.

Let us return to the proof of Theorem 3.2. Using Fourier transform, an equivalent statement of (3.9) is

$$\begin{cases} \sum_i \sum_j \dfrac{a_{ijk}\, \hat{v}_i^\varepsilon(\xi)\, \xi_j}{1 + |\xi|} \to \sum_i \sum_j \dfrac{a_{ijk}\, \hat{v}_i(\xi)\xi_j}{1 + |\xi|} \\ \text{in } L^2(\mathbb{R}^N) \text{ strong, } k = 1,\dots,n. \end{cases} \tag{3.17}$$

Using (3.16) and denoting

$$g_k^\varepsilon(\xi) = \sum_i \sum_j \frac{a_{ijk}\, \hat{v}_i^\varepsilon(\xi)\xi_j}{1 + |\xi|}, \quad g_k(\xi) = \sum_i \sum_j \frac{a_{ijk}\, \hat{v}_i(\xi)\xi_j}{1 + |\xi|}, \tag{3.18}$$

we have:

$$\begin{cases} \displaystyle\int_{|\xi|>R} \tilde{f}(\hat{v}^\varepsilon(\xi))d\xi \geq \\ \displaystyle\geq - \delta \int_{|\xi|>R} |\hat{v}^\varepsilon(\xi)|^2 d\xi - C_\delta \sum_k \int_{|\xi|>R} \Big|\sum_i \sum_j a_{ijk}\, \hat{v}^\varepsilon(\xi) \frac{\xi_j}{|\xi|}\Big|^2 d\xi \geq \\ \displaystyle\geq - \delta \int_{\mathbb{R}^N} |\hat{v}^\varepsilon(\xi)|^2 d\xi - C_\delta \sum_k \int_{|\xi|>R} |g_k^\varepsilon(\xi)|^2 \Big|\frac{1+|\xi|}{|\xi|}\Big|^2 d\xi \geq \\ \displaystyle\geq - \delta M - C_\delta \sum_k \int_{|\xi|>R} |g_k^\varepsilon(\xi)|^2 \Big|\frac{1+|\xi|}{|\xi|}\Big|^2 d\xi, \end{cases} \tag{3.19}$$

where M is a constant independent of ε, R and δ such that:

$$\|v^\varepsilon\|^2_{(L^2(\mathbb{R}^N))^m} \leq M, \quad \forall \varepsilon.$$

Using (3.17) to pass to the limit in (3.19) as $\varepsilon \to 0$ for $R > 0$ and $\delta > 0$ fixed one obtains:

$$\begin{cases} \liminf_{\varepsilon\to 0} \int_{|\xi|>R} \tilde{f}(\hat{v}^{\varepsilon}(\xi))d\xi \geq \\ \geqq -\delta M - C_{\delta} \sum_{k} \int_{|\xi|>R} |g_k(\xi)|^2 \left|\frac{1+|\xi|}{|\xi|}\right|^2 d\xi. \end{cases}$$

For any given $\eta > 0$, a convenient choice of $\delta = \frac{\eta}{2M}$ and then of $R = R_{\eta}$ sufficiently large implies, since g_k belongs to $L^2(\mathbb{R}^N)$ that

$$\liminf_{\varepsilon\to 0} \int_{|\xi|>R} \tilde{f}(\hat{v}^{\varepsilon}(\xi))d\xi \geqq -\eta. \tag{3.20}$$

From (3.12), (3.13), (3.14) and (3.20) it follows that for any $\eta > 0$ fixed

$$\begin{cases} \liminf_{\varepsilon\to 0} \int_{\mathbb{R}^N} f(v^{\varepsilon}(x))dx \geq \\ \geqq \int_{\mathbb{R}^N} f(v(x))dx - \int_{|\xi|>R} \tilde{f}(\hat{v}(\xi))d\xi - \eta \end{cases}$$

which using (3.15) implies (3.11). The proof of Theorem 3.2 is now complete.□

4. A REMARK IN THE CASE WHERE f HAS NON-CONSTANT COEFFICIENTS

In this section we consider a quadratic form f with non-constant coefficients

$$f(x,\lambda) = \sum_{j,j'=1}^{m} f_{jj'}(x) \lambda_j \lambda_{j'} \tag{4.1}$$

where x lies in Ω and λ in $\mathbb{R}^m$. When the coefficients are continuous a weak lower semicontinuity theorem similar to Theorem 3.2 still holds true. At the end of this section we prove, by means of two counterexamples, that the continuity of the coefficients as well as the presence of a smooth function ϕ with compact support are necessary conditions for the functional to be weakly lower semicontinuous.

Theorem 4.1 Let the quadratic form f defined by (4.1) have *continuous* coefficients

$$f_{jj'}(x) \in C^0(\Omega) \tag{4.2}$$

and satisfy

$$f(x,\lambda) \geqq 0 \quad \forall x \in \Omega, \qquad \forall \lambda \in \Lambda . \tag{4.3}$$

Then the functional $u \to f(u)$ is weakly lower semicontinuous in the following sense:

$$\begin{cases} \text{for all sequence } u^{\varepsilon} \text{ satisfying (3.4) and (3.5)} \\ \text{for all } \phi \in \mathcal{D}(\Omega),\ \phi \geqq 0, \\ \text{one has } \liminf\limits_{\varepsilon \to 0} \displaystyle\int_{\Omega} \phi(x)\ f(x,u^{\varepsilon}(x))dx \geqq \int_{\Omega} \phi(x)\ f(x,u(x))dx. \quad \square \end{cases} \tag{4.4}$$

Proof of Theorem 4.1

Let $\delta > 0$ be given. For any multi-index $\ell = (\ell_1,\ell_2,\ldots,\ell_N)$ in $\mathbb{Z}^N$ we define the open cube:

$$C^{\ell} = \{x \in \mathbb{R}^N \mid (\ell_i - 1)\ \delta \leqq x_i \leqq (\ell_i + 1)\delta,\ i = 1,\ldots,N\}$$

of size 2δ and centre x^{ℓ} with

$$x_i^{\ell} = \ell_i,\ i = 1^{\ell},\ldots,N.$$

The cubes C^{ℓ} define an open covering of $\mathbb{R}^N$ and thus there exists a partition of unity which agrees with this covering, i.e. functions ψ_{ℓ} such that

$$\begin{aligned} &\psi^{\ell} \in \mathcal{D}(\mathbb{R}^N),\ 1 \geqq \psi^{\ell} \geqq 0, \\ &\quad \operatorname{supp}\ \psi^{\ell} \subset C^{\ell}, \\ &\sum_{\ell} \psi^{\ell}(x) = 1, \quad \forall x \in \mathbb{R}^N, \end{aligned} \tag{4.5}$$

(these functions can be made from a partition of unity associated with a convenient covering of the N-torus).

Then

$$\begin{cases} \displaystyle\int_\Omega \phi(x)\, f(x,u^\varepsilon(x))dx = \sum_\ell \int_{C^\ell} \psi^\ell(x)\, \phi(x)\, f(x,u^\varepsilon(x))dx = \\ \displaystyle = \sum_\ell \int_{C^\ell} \psi^\ell(x)\, g^\ell(u^\varepsilon(x))dx \quad + \\ \displaystyle + \sum_\ell \int_{C^\ell} \psi^\ell(x)\, [\phi(x)\, f(x,u^\varepsilon(x)) - g^\ell(u^\varepsilon(x))]dx \end{cases} \tag{4.6}$$

where g^ℓ denotes the quadratic form ϕf whose coefficients are frozen at the centre of the cube C^ℓ, i.e.

$$g^\ell(\lambda) = \phi(x^\ell)\, f(x^\ell,\lambda), \quad \forall\lambda \in \mathbb{R}^m.$$

For $\phi \in \mathcal{D}(\Omega)$, $\phi \geqq 0$, the quadratic form g^ℓ satisfies (see (4.3))

$$g^\ell(\lambda) \geqq 0, \ \forall\lambda \in \Lambda. \tag{4.7}$$

On the other hand, since ϕ has the compact support, the index ℓ in (4.6) ranges over a finite set L and

$$\begin{cases} |\phi(x)\, f(x,\lambda) - g^\ell(\lambda)| \leq \omega(\delta)\, |\lambda|^2, \\ \quad \forall\ell \in L,\ \forall x \in C^\ell, \quad \forall\lambda \in \mathbb{R}^m, \end{cases}$$

where $\omega(\delta)$ is the modulus of continuity of ϕf with respect to $x \in \operatorname{supp}\phi$; note that $\omega(\delta) \to 0$ as $\varepsilon \to 0$.

The following estimate of the last term in (4.6) holds true:

$$\begin{cases} \displaystyle |\sum_{\ell\in L} \int_{C^\ell} \psi^\ell(x)[\phi(x)\, f(x,u^\varepsilon(x)) - g^\ell(u^\varepsilon(x))]dx \leq \\ \displaystyle \leqq \sum_{\ell\in L} \int_{C^\ell} \psi^\ell(x)\, \omega(\delta)|u^\varepsilon(x)|^2 dx \leq \omega(\delta)\, \|u^\varepsilon\|^2_{L^2(\Omega)}. \end{cases} \tag{4.8}$$

Consider now a sequence u^ε satisfying (3.4) and (3.5). Theorem 3.2 applied to each quadratic form g^ℓ (in finite number) together with (4.6) and (4.8) implies, for any fixed $\delta > 0$:

$$\begin{cases} \liminf_{\varepsilon \to 0} \int_\Omega \phi(x)\, f(x,u^\varepsilon(x))dx \geq \\ \geq \sum_{\ell \in L} \int_{C^\ell} \psi_\ell(x)\, g^\ell(u(x))dx - \omega(\delta)\ \text{cste} \geq \\ \geq \int_\Omega \phi(x)\, f(x,u(x))dx - \omega(\delta)\, \|u\|^2_{L^2(\Omega)} - \omega(\delta)\ \text{cste}. \end{cases} \tag{4.9}$$

Passing to the limit in (4.9) as δ tends to 0 completes the proof of Theorem 4.1. □

We now prove by means of two counterexamples that the restrictions made in Theorem 4.1 (presence of $\phi \in \mathcal{D}(\Omega)$ in (4.4) and continuity of the coefficients) are essential to get the weak lower semicontinuity.

<u>Example 4.1</u> Let $\Omega = \mathbb{R}^N$, $\tilde{y} \in \mathcal{D}(\mathbb{R}^N)$, $\tilde{w} \in (\mathcal{D}(\mathbb{R}^N))^N$ be such that

$$\int_{\mathbb{R}^N} \tilde{y}\ \text{div}\ \tilde{w} = +1. \tag{4.10}$$

Let us define $u^\varepsilon = (v^\varepsilon, w^\varepsilon) \in (L^2(\mathbb{R}^N))^{2N}$ by:

$$\begin{cases} \tilde{v} = \text{grad}\ \tilde{y}, \\ v^\varepsilon(x) = \tilde{v}(x + \frac{1}{\varepsilon} x_0),\quad w^\varepsilon(x) = \tilde{w}(x + \frac{1}{\varepsilon} x_0), \end{cases}$$

where $x_0 \neq 0$ is a fixed vector in $\mathbb{R}^N$.

The functions v^ε and w^ε are merely translates of $\tilde{v}$ and $\tilde{w}$; thus

$$\begin{cases} v^\varepsilon \longrightarrow v = 0 \text{ in } (L^2(\mathbb{R}^N))^N, \\ w^\varepsilon \longrightarrow w = 0 \text{ in } (L^2(\mathbb{R}^N))^N. \end{cases} \tag{4.11}$$

Since

$$\begin{cases} \text{curl}\ v^\varepsilon (\equiv 0) \text{ is bounded in } (L^2(\mathbb{R}^N))^{N^2}, \\ \text{div}\ w^\varepsilon \text{ is bounded in } L^2(\mathbb{R}^N), \end{cases} \tag{4.12}$$

the sequence u^ε satisfies (3.4) (here (4.11)) and (3.5) (here (4.12)) for the set of coefficients a_{ijk} associated with (4.12). The cone Λ turns out in this case to be (see Example 7.1 below):

$$\Lambda = \{\lambda = (\alpha,\beta) \in \mathbb{R}^{2N} \mid \langle\alpha,\beta\rangle = 0\}$$

where $\langle\cdot,\cdot\rangle$ denotes the inner product in $\mathbb{R}^N$, i.e. where

$$\langle\alpha,\beta\rangle = \sum_{i=1}^{N} \alpha_i\beta_i .$$

Consider now as quadratic form f the inner product $\langle\cdot,\cdot\rangle$, i.e.:

$$f(u) = \langle v,w\rangle .$$

This quadratic form has constant coefficients; it clearly satisfies

$$f(\lambda) = \langle\alpha,\beta\rangle = 0, \quad \forall\lambda = (\alpha,\beta) \in \Lambda .$$

Integration by parts and (4.10) yield:

$$\begin{cases} \displaystyle\int_{\mathbb{R}^N} f(u^\varepsilon(x))dx = \int_{\mathbb{R}^N} \langle v^\varepsilon(x), w^\varepsilon(x)\rangle\, dx = \\ \displaystyle = \int_{\mathbb{R}^N} \langle\tilde{v}(x), \tilde{w}(x)\rangle\, dx = \int_{\mathbb{R}^N} \langle \text{grad}\, \tilde{y}, \tilde{w}\rangle\, dx = \\ \displaystyle = -\int_{\mathbb{R}^N} \tilde{y}\, \text{div}\, \tilde{w}\, dx = -1. \end{cases}$$

On the other hand (see (4.11))

$$\int_{\mathbb{R}^N} f(u(x))dx = \int_{\mathbb{R}^N} f(0)dx = 0.$$

Thus

$$-1 = \liminf_{\varepsilon\to 0} \int_{\mathbb{R}^N} f(u^\varepsilon(x))dx < \int_{\mathbb{R}^N} f(u(x))dx = 0,$$

which proves that, in Theorem 4.1, the presence of a function ϕ with compact support is necessary for (4.4) to hold true. □

Remark 4.1 The previous Example 4.1 comes from the classical idea of shifting the mass at infinity, often used to obtain counterexamples in integration theory. It works in the present context because there exists a quadratic form (the inner product $\langle\cdot,\cdot\rangle$) which is weakly continuous under the conditions (4.11), (4.12). (This is just the "div-curl lemma", see Example 7.1 below). The unboundedness of Ω $(= \mathbb{R}^N)$ is of course essential here. □

Remark 4.2 The quadratic form of Example 4.1 has constant coefficients. Thus this example also applies to the situation of Theorem 3.2 and Remark 3.2, showing the necessity of the presence of a test function ϕ with compact support. □

Example 4.2 Set $Q =]-1, +1[^2$. Consider y^ε and u^ε defined on Q by:

$$\begin{cases} y_1^\varepsilon(x_1,x_2) = \sqrt{\varepsilon}\,(1 - |x_2|)^{1/\varepsilon} \cos \dfrac{x_1}{\varepsilon}, \\ y_2^\varepsilon(x_1,x_2) = \sqrt{\varepsilon}\,(1 - |x_2|)^{1/\varepsilon} \sin \dfrac{x_1}{\varepsilon}, \\ u_{ij}^\varepsilon = \dfrac{\partial y_i^\varepsilon}{\partial x_j}, \qquad i,j = 1,2. \end{cases}$$

Then

$$\begin{cases} \|u^\varepsilon\|^2_{(L^2(Q))^4} = \dfrac{2}{\varepsilon} \displaystyle\int_{-1}^{+1} dx_1 \int_0^1 \{|1-x_2|^{2/\varepsilon} + |1-x_2|^{(2/\varepsilon)-2}\} dx_2 = \\ \qquad = \dfrac{4}{\varepsilon} \left\{ \dfrac{1}{\frac{2}{\varepsilon} + 1} + \dfrac{1}{\frac{2}{\varepsilon} - 1} \right\} < 8 . \end{cases}$$

Since $\|y^\varepsilon\|_{(L^\infty(Q))^2} \leq \sqrt{\varepsilon}$, it follows:

$$u^\varepsilon \longrightarrow u = 0 \text{ in } (L^2(Q))^4 \text{ weak.} \tag{4.13}$$

Since

$$\frac{\partial u_{ij}^\varepsilon}{\partial x_k} - \frac{\partial u_{ik}^\varepsilon}{\partial x_j} = 0 \quad i,j,k = 1,2, \tag{4.14}$$

the sequence u^ε satisfies (3.4) (here (4.13)) and (3.5) (here (4.14)), for any open set $\Omega \subset Q$.

In this case, the cone Λ turns out to be (see Example 7.2 below)

$$\Lambda = \{\lambda \in \mathbb{R}^4 | \exists \alpha \in \mathbb{R}^2, \exists \xi \in \mathbb{R}^2 - \{0\},$$

$$\lambda_{ij} = \alpha_i \xi_j \quad i,j = 1,2\}.$$

Consider as quadratic form f the determinant defined by:

$$f(\lambda) = \det(\lambda) = \lambda_{11} \lambda_{22} - \lambda_{12} \lambda_{21},$$

which has constant coefficients and satisfies

$$f(\lambda) = 0, \forall \lambda \in \Lambda.$$

For $0 < a < 1$ let Ω_a be defined by

$$\Omega_a =]-a, + a[\times]0,a[\subset Q.$$

Then

$$\int_{\Omega_a} \det(u)dx = \int_{\Omega_a} \det(0)dx = 0$$

and

$$\begin{cases} \displaystyle\int_{\Omega_a} \det(u^\varepsilon)dx = \int_{-a}^{+a} dx_1 \int_0^a (-\frac{1}{\varepsilon}) (1 - x_2)^{\frac{2}{\varepsilon} - 1} dx_2 = \\ \qquad = a[(1-a)^{2/\varepsilon} - 1]. \end{cases}$$

So we obtain:

$$- a = \lim_{\varepsilon \to 0} \int_{\Omega_a} \det(u^\varepsilon)dx < \int_{\Omega_a} \det(u)dx = 0. \qquad (4.15)$$

This example has to be contrasted with Theorem 4.1. This can be done in three (more or less parallel) ways: first take $\Omega = \Omega_a$ and $\phi = 1$ in (4.4); then (4.15) proves the necessity of a test function with compact support, even if Ω is bounded. Second, take $\Omega = Q$ and ϕ to be the characteristic function of Ω_a in (4.4): then (4.15) proves that the test function ϕ has to be smooth (at least continuous). Finally take $\Omega = Q$, $\phi \in \mathcal{D}(\Omega)$, $\phi = 1$ on Ω_a and $f(x,u)$ with non constant, discontinuous coefficients

$$\begin{cases} f(x,u) = \det(u) & \text{if } x \in \Omega_a, \\ f(x,u) = 0 & \text{if } x \in Q \setminus \Omega_a, \end{cases}$$

which satisfies (4.3); then (4.15) proves the necessity of a continuous dependence of f upon x. □

Remark 4.3 Example 4.2 is precisely Counterexample 7.3 of my joint paper [8] with John Ball. This counterexample proves that $y \to \det \nabla y$ is not continuous from $W^{1,N}(\mathbb{R}^N;\mathbb{R}^N)$ endowed with its weak topology into $L^1(\mathbb{R}^N)$ endowed with its *weak* topology (here N = 2). An analogous example, related to the "div-curl lemma", can be found in [6] pp. 252-253. □

5. FURTHER NECESSARY CONDITIONS FOR WEAK CONTINUITY

In Section 2, we gave a first necessary condition for the weak lower semicontinuity (Theorem 2.1) and for the weak continuity (Theorem 2.2) of the mapping $u \to f(u)$. In this section we state further conditions for weak continuity. For the details we refer to [3] pp. 186-190 and [7] Section 2.

If $u \to f(u)$ is weakly continuous, we know from Theorem 2.1 that f is affine in the directions of Λ. Thus the derivatives of f in these directions make sense.

The idea of the proof of Theorem 2.1 was to consider the sequence u^ε defined by

$$u^\varepsilon(x) = \chi(\frac{x\cdot\xi}{\varepsilon})\, a + (1-\chi(\frac{x\cdot\xi}{\varepsilon}))b \tag{5.1}$$

where $((a-b),\xi)$ belongs to V defined by (1.9) and where χ is a periodic characteristic step function. In this case it is easy to compute the weak limit of u^ε and of any $f(u^\varepsilon)$.

The idea here is to add functions of the type (5.1), with several oscillating directions ξ. When these directions are linearly dependent, "interactions" between the oscillations occur, yielding a new necessary condition. Using $r \geq 2$ directions, one obtains (cf. [3], Theorem 18 and

[7], Proposition 2.6):

Theorem 5.1 If the mapping $u \to f(u)$ is weakly continuous from Z_∞ defined by (1.5) into $L^\infty(\Omega)$, then for all integers $r \geq 2$, the function f necessarily satisfies:

$$
\begin{cases}
\forall(\lambda_1,\xi_1), (\lambda_2,\xi_2),\ldots,(\lambda_n,\xi_r) \in V \\
\text{with } \xi_1,\xi_2,\ldots,\xi_r \text{ linearly dependent,} \\
\forall a \in \mathbb{R}^m, \\
f^{(r)}(a)\ \lambda_1\lambda_2 \ldots\lambda_r = 0. \quad \square
\end{cases}
\tag{5.2}
$$

Remark 5.1 It can be proved (see [3] pp. 185-186 or [6] p. 248) that at least in some cases (depending of course on the specific form of the cone Λ), condition (5.2) with $r > 2$ is more stringent that the first necessary condition stated in Theorem 2.2 which merely corresponds to $r = 2$. □

Let us denote by L the linear subspace of $\mathbb{R}^m$ spanned by the cone Λ defined by (1.10); let ℓ be its dimension ($0 \leq \ell \leq m$). At the expense of a possible change of basis in $\mathbb{R}^m$ (the new basis need not be orthogonal) we can assume that L coincides with the subspace described by the first ℓ components $(a_1,a_2,\ldots,a_\ell)$ of $a \in \mathbb{R}^m$. From Theorem 5.1 a further analysis yields (see [7], Section 2):

Theorem 5.2 Let us assume that the mapping $u \to f(u)$ is weakly continuous from Z_∞ into $L^\infty(\Omega)$ in the sense of Question 1 (Section 1). Then f is a finite sum of polynomials P_α acting on $(a_1,\ldots,a_\ell)$; each P_α is homogeneous with constant coefficients and degree d_α less than or equal to $\inf(N,\ell)$; the coefficients of the sum are continuous functions c_α acting on $(a_{\ell+1},\ldots,a_m)$: i.e.

$$
f(a) = \sum_{\substack{\alpha \\ \text{finite}}} c_\alpha(a_{\ell+1},\ldots,a_m)\ P_\alpha(a_1,\ldots,a_\ell). \tag{5.3}
$$

Moreover, each polynomial P_α satisfies (5.2), for every integer $r \geq 2$ (or equivalently for r equal to the degree d_α of P_α if $d_\alpha \geq 2$). □

6. THESE NECESSARY CONDITIONS ARE SUFFICIENT UNDER A TECHNICAL ASSUMPTION

Theorem 5.2 gives necessary conditions on f for the mapping $u \to f(u)$ to be weakly continuous. In this section we state that these necessary conditions are also sufficient for the weak continuity if a technical hypothesis is met by the coefficients a_{ijk}; for the details we refer to [7], Section 3.

Theorem 5.2 asserts that the weak continuity of the mapping $u \to f(u)$ implies that f is of the form (5.3). Let us first prove that the coefficients c_α (in the notation of Theorem 5.2) behave nicely.

Proposition 6.1 For any continuous function $c_\alpha : \mathbb{R}^{m-\ell} \to \mathbb{R}$, the mapping $u \to c_\alpha(u) = c_\alpha(u_{\ell+1}, \ldots\ldots, u_m)$ is sequentially continuous from Z_∞ (endowed with its *weak* $\star$ topology) into $L^p(\Omega)$ (endowed with its *strong* topology), i.e.:

$$\begin{cases} c^\alpha(u^\varepsilon_{\ell+1}, \ldots, u^\varepsilon_m) \to c^\alpha(u_{\ell+1}, \ldots, u_m) \text{ in } L^p(\Omega) \text{ strong } \forall p < +\infty, \\ \text{for any sequence } u^\varepsilon \text{ satisfying (1.1) and (1.2).} \end{cases} \quad (6.1)$$

□

Proof Let us consider a sequence u^ε satisfying (1.1) and (1.2), and the quadratic form defined on $\mathbb{R}^m$ by:

$$f_j(\lambda) = \lambda_j^2$$

with $\ell+1 \leq j \leq m$. Because of the definition of L and Λ, we have $\Lambda \subset L$ and

$$f_j(\lambda) = 0, \quad \forall \lambda \in \Lambda.$$

Theorem 3.3 implies that f_j is weakly continuous. Thus the sequence u^ε satisfies

$$(u_j^\varepsilon)^2 \longrightarrow (u_j)^2 \text{ in } L^\infty(\Omega) \text{ weak } \star.$$

This implies in turn the strong convergence of $u_j^\varepsilon (\ell+1 \leq j \leq m)$ in $L^2(\Omega)$. Proposition 6.1 is proved. □

Thus in view of (5.3) the problem is reduced to the study of the weak continuity of the mappings $u \to P_\alpha(u_1, \ldots, u_\ell)$ where P_α is a polynomial satisfying (5.2) for every $r \geq 2$. If we succeed in proving that such polynomials generate weakly continuous mappings, the necessary conditions of Theorem 5.2 also become sufficient.

This result is already known (Theorem 3.1) when the degree of the polynomial P_α is 2. When the degree is greater or equal to 3, the result still holds true (at least under a technical hypothesis) because of

Theorem 6.1 Let us assume that the coefficients a_{ijk} are such that:

$$\begin{cases} \text{for every } \xi \in \mathbb{R}^N - \{0\}, \\ \text{rank } \{\lambda \in \mathbb{R}^m \mid (\lambda,\xi) \in V\} \text{ is independent of } \xi \end{cases} \quad . \tag{6.2}$$

If P_α denotes a homogeneous polynomial with constant coefficients acting on $(a_1,\ldots,a_\ell)$ and satisfying (5.2) for every integer $r \geq 2$ (or equivalently for r equal to the degree of P_α), the mapping $u \to P_\alpha(u)$ is weakly continuous in the sense of Question 1 (Section 1). □

Thus under the hypothesis (6.2) the necessary conditions given in Theorem 5.2 for $u \to f(u)$ to be weakly continuous are also sufficient. In this case, the weakly continuous mappings on Z_∞ are completely characterized by means of a finite number of the algebraic relations expressed in (5.2).

Let us sketch the proof of Theorem 6.1 given in [7] and explain why the hypothesis (6.2) is required in this proof. As in the proof of Theorems 3.1 and 3.2, the first step is to localize and to reduce the problem to the case where the test function ϕ is 1 and the sequence u^ε under consideration is defined on the whole $\mathbb{R}^N$, with fixed compact support K. The second step consists of decomposing u^ε in two parts. Using the Fourier transform, we define:

$$\begin{cases} u^\varepsilon = v^\varepsilon + w^\varepsilon, \\ \hat{v}^\varepsilon(\xi) = \text{Proj}_{K(\xi)}\, \hat{u}^\varepsilon(\xi), \\ \hat{w}^\varepsilon(\xi) = \text{Proj}_{(K(\xi))^\perp}\, \hat{u}^\varepsilon(\xi), \\ K(\xi) = \{\lambda \in \mathbb{R}^m \mid (\lambda,\xi) \in V\}. \end{cases} \tag{6.3}$$

The constant rank hypothesis (6.2) is crucial here; it allows us to use Mikhlin's theorem on Fourier multipliers and ensures that v^ε and w^ε are bounded in $(L^p(\mathbb{R}^N))^m$ for all p $(1 < p < +\infty)$. Moreover the constant rank hypothesis also implies that w^ε is bounded on $(H^1(\mathbb{R}^N))^m$, by virtue of a result due to J.R. Schulenberger and C.H. Wilcox.

The third step is to pass to the limit. To this effect, the polynomial character of P_α is used to expand

$$\int_{\mathbb{R}^N} P_\alpha(u^\varepsilon(x))dx = \int_{\mathbb{R}^N} P_\alpha(v^\varepsilon(x) + w^\varepsilon(x))dx$$

in a sum of terms. The term $\int_{\mathbb{R}^N} P_\alpha(v^\varepsilon(x))dx$ is the most delicate one. Using Plancherel-Parseval identity and $(v_i^\varepsilon v_j^\varepsilon)\hat{} = \hat{v}_i^\varepsilon * \hat{v}_j^\varepsilon$ this term becomes a convolution product on the Fourier transforms $\hat{v}_i^\varepsilon$. But $(\hat{v}^\varepsilon(\xi),\xi) \in V$ by the definition (6.3) of v^ε, and the integrand turns out to be equal to zero since the necessary condition (5.2) is satisfied. All the other terms of the sum contain at least one w_i^ε. Using again the Plancherel-Parseval identity, they are expressed as sums (in i) of terms of the type

$$\int_{\mathbb{R}^N} \hat{w}_i^\varepsilon(\xi)\overline{(Q_i(v^\varepsilon,w^\varepsilon))}\hat{}(\xi)d\xi$$

where Q_i is a polynomial of degree $d_\alpha - 1$ (P_α has degree d_α). As in (3.12), these terms are decomposed into two parts:

$$\int_{|\xi|\leq R} + \int_{|\xi|>R} .$$

The term at infinity is controlled using the boundedness of w^ε in $(H^1(\mathbb{R}^N))^m$. Passing to the limit in the term $\int_{|\xi|\leq R}$ uses the weak convergence of $Q_i(v^\varepsilon,w^\varepsilon)$ to $Q_i(v,w)$, a result which follows from the fact that Q_i also satisfies (5.2) and is of lower degree (the proof is actually an induction process on the degree of the polynomial).

In this proof, the necessity of the assumption (6.2) is due to the use of the decomposition (6.3). Such a decomposition was not used in the proof of Theorem 3.2, where the terms at infinity were globally estimated by means of Lemma 3.1, but the proof here is more "primitive".

We believe that hypothesis (6.2) is superfluous. Indeed the result of Theorem 6.1 can be obtained in cases where hypothesis (6.2) is violated (see example 7.3 below).

7. EXAMPLES

7.1 The "div-curl" Lemma

This is actually the prototype of the whole compensated compactness. This result is one of the basic tools in homogenization theory. It is also one of the roots of the method of compensated compactness applied to systems of hyperbolic equations (see [3] Section 7, and [4]).

Theorem 7.1 Let Ω be an open set of $\mathbb{R}^N$ and v^ε and w^ε be two sequences such that:

$$\begin{cases} v^\varepsilon \longrightarrow v \text{ in } (L^2(\Omega))^N \text{ weak}, \\ w^\varepsilon \longrightarrow w \text{ in } (L^2(\Omega))^N \text{ weak}, \end{cases} \tag{7.1}$$

$$\begin{cases} \text{curl } v^\varepsilon \text{ bounded in } (L^2(\Omega))^{N^2} \text{ (or compact in } (H^{-1}(\Omega))^{N^2}), \\ \text{div } w^\varepsilon \text{ bounded in } L^2(\Omega) \text{ (or compact in } H^{-1}(\Omega)). \end{cases} \tag{7.2}$$

Let $\langle\cdot,\cdot\rangle$ denote the inner product in $\mathbb{R}^N$, i.e.:

$$\langle v,w\rangle = \sum_{i=1}^{N} v_i w_i .$$

Then

$$\langle v^\varepsilon,w^\varepsilon\rangle \longrightarrow \langle v,w\rangle \text{ in } \mathcal{D}'(\Omega). \tag{7.3}$$

Moreover, the inner product $\langle\cdot,\cdot\rangle$ is the only non linear function f which generates a weakly continuous mapping $(v,w) \to f(v,w)$ (in the sense of question 1) for sequences satisfying (7.1), (7.2). □

For any $v \in (L^2(\Omega))^N$, curl v is defined by:

$$(\text{curl } v)_{ij} = \frac{\partial v_i}{\partial x_j} - \frac{\partial v_j}{\partial x_i} \quad i,j = 1,\dots,N. \tag{7.4}$$

Thus the manifold V and the cone Λ associated with the condition (7.2) are:

$$\begin{cases} V = \{(\lambda,\xi) | \lambda = (\alpha,\beta) \in \mathbb{R}^N \times \mathbb{R}^N, \ \xi \in \mathbb{R}^N - \{0\}, \\ \qquad\qquad \xi_j\alpha_i - \xi_i\alpha_j = 0, i,j = 1,\ldots,N, \ \sum_{i=1}^N \xi_i\beta_i = 0\}, \\ \Lambda = \{\lambda | \lambda = (\alpha,\beta) \in \mathbb{R}^N \times \mathbb{R}^N, \ \langle\alpha,\beta\rangle = \sum_{i=1}^N \alpha_i\beta_i = 0\}. \end{cases} \tag{7.5}$$

Proof of Theorem 7.1

The inner product $\langle\cdot,\cdot\rangle$ is a quadratic form which is clearly zero on Λ. Thus (7.3) follows from Theorem 3.3, or more exactly from a slightly stronger version similar to Theorem 3.2 (see Remark 3.2).

Conversely, if f is a function such that the mapping $(v,w) \to f(v,w)$ is weakly continuous, we know from Theorem 2.2 that

$$t \in \mathbb{R} \to f(a + t\alpha, b + t\beta) \in \mathbb{R}$$

is affine for all $(a,b) \in \mathbb{R}^N \times \mathbb{R}^N$ and all $(\alpha,\beta) \in \Lambda$. Taking $\alpha = 0$ then $\beta = 0$ this implies:

$$f(a,b) = f_o + \sum_{i=1}^N (A_i a_i + B_i b_i) + \sum_{i,j=1}^N f_{ij} a_i b_j .$$

But the quadratic part has to be zero if $\langle a,b\rangle = 0$. Thus

$$f(a,b) = f_o + \sum_{i=1}^N (A_i a_i + B_i b_i) + c \sum_{i=1}^N a_i b_i .$$

The proof of the theorem is complete. □

Remark 7.1 There is another proof of (7.3). Indeed after localization and extension by zero of the functions v^ε and w^ε to all of $\mathbb{R}^N$, one can decompose v^ε, setting

$$v^\varepsilon = z^\varepsilon + \text{grad } y^\varepsilon,$$

where z^ε is defined by:

$$\text{curl } z^\varepsilon = \text{curl } v^\varepsilon, \quad \text{div } z^\varepsilon = 0 \text{ in } \mathbb{R}^N.$$

The term z^ε turns out to be bounded on $(H^1(\mathbb{R}^N))^N$ (see [5], Lemma 2), thus compact in $(L^2_{loc}(\mathbb{R}^N))^N$. Similarly, since grad $y^\varepsilon = v^\varepsilon - z^\varepsilon$ is bounded in

$(L^2(\mathbb{R}^N))^N$, y^ε is compact in $L^2_{loc}(\mathbb{R}^N)$. An integration by parts performed on a bounded open set ω (recall that after localization v^ε and w^ε have compact supports included in some fixed compact K) yields:

$$\begin{cases} \int_{\mathbb{R}^N} \langle v^\varepsilon, w^\varepsilon \rangle dx = \int_\omega \langle v^\varepsilon, w^\varepsilon \rangle dx = \\ = \int_\omega \langle z^\varepsilon, w^\varepsilon \rangle dx - \int_\omega y^\varepsilon \operatorname{div} v^\varepsilon. \end{cases} \tag{7.6}$$

Using the compactness of z^ε and y^ε in L^2_{loc}, it is easy to pass to the limit in each term of (7.6). Another integration by parts gives:

$$\int_{\mathbb{R}^N} \langle v^\varepsilon, w^\varepsilon \rangle dx \to \int_{\mathbb{R}^N} \langle v, w \rangle dx.$$

Thus (7.3) is (once again) proved.

This proof also works in the case where L^2 is replaced in (7.1), (7.2) by L^p spaces: L^p for v^ε and curl v^ε, $L^{p'}$ for w^ε and div w^ε, with $\frac{1}{p} + \frac{1}{p'} = 1$: for the details, see [5], Section 2. □

Remark 7.2 In (7.3) the convergence takes place in $\mathcal{D}'(\Omega)$, and (since $\langle v^\varepsilon, w^\varepsilon \rangle$ is bounded in $L^1(\Omega)$) in the vague topology of measures. But the convergence does not take place in the *weak* topology of $L^1(\Omega)$ (see [6], pp. 252-253 for a counterexample). For a (more simple) example showing the necessity of the presence of a test function ϕ with compact support when $\Omega = \mathbb{R}^N$, see example 4.1 above. □

7.2 The case of gradients

This is a very important example, for which C.B. Morrey defined the notion of quasiconvexity (see [9]).

Theorem 7.2 Let Ω be an open bounded set of $\mathbb{R}^N$ and y^ε be a sequence such that:

$$y^\varepsilon \rightharpoonup y \text{ in } (W^{1,\infty}(\Omega))^q \text{ weak } \star. \tag{7.7}$$

Denote by ∇y^ε the $q \times N$ matrix of coefficients

$$(\nabla y^\varepsilon)_{ij} = \frac{\partial y_i^\varepsilon}{\partial x_j} \tag{7.8}$$

and by sub det (M) the determinant of some $s \times s$ $(1 \leqq s \leqq \inf(N,q))$ submatrix of M. Then

$$\text{sub det}\,(\nabla y^{\varepsilon}) \longrightarrow \text{sub det}\,(\nabla y) \text{ in } L^{\infty}(\Omega) \text{ weak } *. \tag{7.9}$$

Conversely, let f be a continuous function from $\mathbb{R}^{q\times N}$ into $\mathbb{R}$. If

$$f(\nabla y^{\varepsilon}) \longrightarrow f(\nabla y) \text{ in } L^{\infty}(\Omega) \text{ weak } * \tag{7.10}$$

for all sequences y^{ε} satisfying (7.7), then f(M) is a sum (with constant coefficients) of determinants of sub matrices of the $q \times N$ matrix M. □

Theorem 7.3 In the setting of Theorem 7.2, let f be a continuous function from $\mathbb{R}^{q \times N}$ into $\mathbb{R}$ such that

$$\begin{cases} \forall \phi \in \mathcal{D}(\Omega), \quad \phi \geqq 0, \\ \liminf\limits_{\varepsilon \to 0} \displaystyle\int_{\Omega} \phi(x) f(\nabla y^{\varepsilon}(x))dx \geqq \int_{\Omega} \phi(x) f(\nabla y(x))dx \\ \text{for all sequences } y^{\varepsilon} \text{ satisfying (7.7)}. \end{cases} \tag{7.11}$$

Then necessarily:

$$\begin{cases} \forall M \in \mathbb{R}^{q\times N}, \quad \forall a \in \mathbb{R}^{q}, \quad \forall b \in \mathbb{R}^{N}, \\ t \in \mathbb{R} \to f(M + t\, a \otimes b) \text{ is convex}, \end{cases} \tag{7.12}$$

where $a \otimes b$ is the rank-1 matrix defined by:

$$(a \otimes b)_{ij} = a_i b_j, \; i = 1,\dots,q, \; j = 1,\dots,N. \tag{7.13}$$

Conversely, when f is quadratic,(7.11) follows from (7.12). □

Let us first explain how this example falls within the scope of the general setting of Section 1. Since the ith line $(1 \leqq i \leqq q)$ of the matrix ∇y^{ε} defined by (7.8) is grad y_i^{ε}, its curl (defined by (7.4)) is zero. Thus we can also consider the sequences u^{ε} such that

$$\begin{cases} u^\varepsilon \longrightarrow u \text{ in } (L^\infty(\Omega))^{q\times N} \text{ weak } \star , \\ \dfrac{\partial u^\varepsilon_{ij}}{\partial x_k} - \dfrac{\partial u^\varepsilon_{ik}}{\partial x_j} \text{ bounded in } L^2(\Omega), \\ \text{for } i = 1,\ldots,q \text{ and } j,k = 1,\ldots,N. \end{cases} \tag{7.14}$$

Actually Theorems 7.2 and 7.3 still hold true when the sequence ∇y^ε is replaced by a sequence u^ε satisfying (7.14).

The manifold V and the cone Λ associated to (7.14) are:

$$\begin{aligned} V = \{(\lambda,\xi) \mid \lambda \in \mathbb{R}^{q\times N},\ \xi \in \mathbb{R}^N - \{0\}, \\ \lambda_{ij}\,\xi_k - \lambda_{ik}\,\xi_j = 0,\ \ i = 1,\ldots,q,\ j,\ k = 1,\ldots,N\} = \\ = \{(\lambda,\xi) \mid \xi \in \mathbb{R}^N - \{0\},\ \ \exists a \in \mathbb{R}^q,\ \lambda = a \otimes \xi\}. \end{aligned} \tag{7.15}$$

$$\Lambda = \{\lambda \mid \lambda \in \mathbb{R}^{q\times N},\ \ \exists a \in \mathbb{R}^q,\ \ \exists \xi \in \mathbb{R}^N - \{0\},\ \lambda = a \otimes \xi\}.$$

Remark 7.3 Theorem 7.2 is concerned with the sequentially weak continuity of the mapping $y \to f(\nabla y)$; it gives a complete answer to this problem.

Theorem 7.3 is concerned with the weak lower semicontinuity of the associated functional; it gives a necessary condition (7.12) which turns out to be sufficient when f is quadratic. When f is not quadratic, the problem of giving an "algebraic" characterization of the possible f is open. Actually a characterization was given by C.B. Morrey [9]: when f is a continuous finite function he proved that the functional $y \to \int_\Omega \phi(x) f(\nabla y(x)dx$ is weakly lower semicontinuous (for y in $(W^{1,\infty}(\Omega))^q$) if and only if f satisfies the following condition, called by him "*quasiconvexity*" in [9] (Note he later changed his terminology)

$$\begin{cases} \forall M \in \mathbb{R}^{q\times N},\ \forall y \in (\mathcal{D}(\Omega))^q, \\ \displaystyle\int_\Omega f(M + \nabla y(x))dx \geq \int_\Omega f(M)dx. \end{cases} \tag{7.16}$$

Unfortunately, the quasi-convexity condition (7.16) uses an infinite number of test functions y and thus is by no means an easy condition to handle. In contrast, the necessary condition (7.12) is more "algebraic"; when f is a C^2 function, (7.12) is equivalent to

$$\begin{cases} \forall M \in \mathbb{R}^{q\times N}, \ \forall a \in \mathbb{R}^q, \ \forall b \in \mathbb{R}^N, \\ f''(M)\, a \otimes b \ \ a \otimes b \geq 0, \end{cases} \tag{7.17}$$

a condition known as the "*Legendre-Hadamard condition*" or the "*rank-one convexity*". Whether this condition is sufficient or not to ensure the weak lower semicontinuity of the functional is an important problem, still open since C.B. Morrey's original work [9].

Note that condition (7.17) reduces to the usual convexity when either q or N is equal to 1; in the other cases, the convexity and rank-one convexity are quite different (compare the rank-one matrices $a \otimes b$ to $\mathbb{R}^{q\times N}$) .

An important class of functions f generating weakly lower semicontinuous functionals was introduced by J.M. Ball and called by him "*polyconvex functions*" (see [10]): this is the class of the functions f of M which can be written as a convex function g acting on the determinants of all the submatrices of M, roughly speaking:

$$f(M) = g(\text{sub det } (M)) \text{ with } g \text{ convex.}$$

Due to the weak continuity of the determinants, it is easy to prove that polyconvex functions generate weakly lower semicontinuous functionals. Since they allow us to prove existence results, and since some examples are found to fit known mechanical models compatible with the requirements of the non linear elasticity theory, these functions play an important role in the applications (see [10]). □

<u>Proof of Theorems 7.2 and 7.3</u>

In view of (7.15), Theorem 7.3 follows immediately from Theorems 2.1 and 3.1; note that in the proof of Theorem 2.1, we only used sequences u^ε such that $\sum_i \sum_j a_{ijk} \frac{\partial u_i^\varepsilon}{\partial x_j} = 0$.

To prove Theorem 7.2, note that all of $\mathbb{R}^{q\times N}$ is spanned by Λ (i.e. $\ell = m = q \times N$ in the notation of Theorems 5.2 and 6.1) and that:

$$\text{rank } \{\lambda | (\lambda,\xi) \in V\} = q \text{ is independent of } \xi. \tag{7.18}$$

Thus hypothesis (6.2) is satisfied in the present case, and Theorems 5.1 and 6.1 apply. The problem reduces to identifying the polynomials P acting on

$\mathbb{R}^{q\times N}$, homogeneous, of degree $d \leq N$, with constant coefficients and such that condition (5.2) is satisfied for $r = d$, i.e. satisfying

$$\begin{cases} P^{(d)}(a_1 \otimes \xi_1)(a_2 \otimes \xi_2) \ldots (a_d \otimes \xi_d) = 0, \\ \forall a_1,\ldots,a_d \in \mathbb{R}^q, \\ \forall \xi_1,\ldots,\xi_d \in \mathbb{R}^N, \\ \text{with } \xi_1,\ldots,\xi_d \text{ linearly dependent.} \end{cases} \tag{7.19}$$

These P turn out to be the sum of determinants of $d \times d$ submatrices; this results from elementary (but tedious) algebraic computations. For the case $N = q = 3$, one can see [10], proof of Theorem 4.1; note this proof only uses condition (2.3) and not the more precise condition (5.2) (here (7.19)). Thus in the particular case studied in this subsection, the "further necessary conditions" (5.2) actually result from the "first necessary condition" (2.3). □

Remark 7.4 There is another proof of Theorem 7.2 due to Y.G. Reshetnyak ([11] Theorem 4) and J.M. Ball([10], Lemma 6.1 and Corollary 6.2.2).

Firstly consider the case of one 2×2 determinant, say for simplicity:

$$\text{sub}_2\det(\nabla y^\varepsilon) = \frac{\partial y_1^\varepsilon}{\partial x_1}\frac{\partial y_2^\varepsilon}{\partial x_2} - \frac{\partial y_1^\varepsilon}{\partial x_2}\frac{\partial y_2^\varepsilon}{\partial x_1} \quad . \tag{7.20}$$

When $y^\varepsilon \in (W^{1,\infty}(\Omega))^q$, one has:

$$\text{sub}_2\det(\nabla y^\varepsilon) = \frac{\partial}{\partial x_1}\left(y_1^\varepsilon \frac{\partial y_2^\varepsilon}{\partial x_2}\right) - \frac{\partial}{\partial x_2}\left(y_1^\varepsilon \frac{\partial y_2^\varepsilon}{\partial x_1}\right) \tag{7.21}$$

in the sense of distributions. But using (7.7) y^ε converges in $(L^\infty(\Omega))^q$ strongly, and it is easy to pass to the limit in (7.21), still in the sense of distributions. This proof is valid for any 2×2 subdeterminant and extends to the subdeterminants of higher order. For example for the subdeterminant associated with the 3×3 submatrix $\frac{\partial y_i}{\partial x_j}$ $(1 \leq i, j \leq 3)$ one has:

$$\text{sub}_3\det \nabla y^{\varepsilon} = \frac{\partial y_1^{\varepsilon}}{\partial x_1}\left[\frac{\partial y_2^{\varepsilon}}{\partial x_2}\frac{\partial y_3^{\varepsilon}}{\partial x_3} - \frac{\partial y_2^{\varepsilon}}{\partial x_3}\frac{\partial y_3^{\varepsilon}}{\partial x_2}\right]$$

$$- \frac{\partial y_1^{\varepsilon}}{\partial x_2}\left[\frac{\partial y_2^{\varepsilon}}{\partial x_1}\frac{\partial y_3^{\varepsilon}}{\partial x_3} - \frac{\partial y_2^{\varepsilon}}{\partial x_3}\frac{\partial y_3^{\varepsilon}}{\partial x_1}\right]$$

$$+ \frac{\partial y_1^{\varepsilon}}{\partial x_3}\left[\frac{\partial y_2^{\varepsilon}}{\partial x_1}\frac{\partial y_3^{\varepsilon}}{\partial x_2} - \frac{\partial y_2^{\varepsilon}}{\partial x_2}\frac{\partial y_3^{\varepsilon}}{\partial x_1}\right] =$$

$$= \frac{\partial}{\partial x_1}\left\{y_1^{\varepsilon}\left[\frac{\partial y_2^{\varepsilon}}{\partial x_2}\frac{\partial y_3^{\varepsilon}}{\partial x_3} - \frac{\partial y_2^{\varepsilon}}{\partial x_3}\frac{\partial y_3^{\varepsilon}}{\partial x_2}\right]\right\} \qquad (7.22)$$

$$- \frac{\partial}{\partial x_2}\left\{y_1^{\varepsilon}\left[\frac{\partial y_2^{\varepsilon}}{\partial x_1}\frac{\partial y_3^{\varepsilon}}{\partial x_3} - \frac{\partial y_2^{\varepsilon}}{\partial x_3}\frac{\partial y_3^{\varepsilon}}{\partial x_1}\right]\right\}$$

$$+ \frac{\partial}{\partial x_3}\left\{y_1^{\varepsilon}\left[\frac{\partial y_2^{\varepsilon}}{\partial x_1}\frac{\partial y_3^{\varepsilon}}{\partial x_2} - \frac{\partial y_2^{\varepsilon}}{\partial x_2}\frac{\partial y_3^{\varepsilon}}{\partial x_1}\right]\right\} .$$

But y_1^{ε} converges in $L^{\infty}(\Omega)$ strongly and we already know that the 2×2 determinants are weakly continuous (in the sense of distributions, but also in $L^{\infty}(\Omega)$ weak $\star$, because they are bounded in $L^{\infty}(\Omega)$). Thus to pass to the limit (in the sense of distributions) in the last expression of (7.22) is easy.

The previous proof relies on the div-curl lemma (see subsection 7.1 above) in the following way: in (7.22) the second expression reads as

$$\sum_{j=1}^{3} \frac{\partial y_1^{\varepsilon}}{\partial x_j} d_{1j}(y^{\varepsilon})$$

where $d_{1j}(y^{\varepsilon})$ is defined as the cofactor of $\dfrac{\partial y_1^{\varepsilon}}{\partial x_j}$. But

$$\frac{\partial y_1^{\varepsilon}}{\partial x_j} \longrightarrow \frac{\partial y_1}{\partial x_j} \text{ in } L^{\infty}(\Omega) \text{ weak } \star,\ j = 1,2,3 ,$$

and

$$\frac{\partial}{\partial x_k}\left(\frac{\partial y_1^{\varepsilon}}{\partial y_j}\right) - \frac{\partial}{\partial x_j}\left(\frac{\partial y_1^{\varepsilon}}{\partial x_k}\right) = 0 ,\ j,k = 1,2,3.$$

On the other hand from the study of the 2×2 determinants, we know that:

$$d_{1j}(y^{\varepsilon}) \longrightarrow d_{1j}(y) \text{ in } L^{\infty}(\Omega) \text{ weak } \star, \ j = 1,2,3,$$

and a computation proves that

$$\sum_{j=1}^{3} \frac{\partial}{\partial x_j} d_{1j}(y^{\varepsilon}) = 0.$$

This is indeed the very proof of the second equality in (7.22). Thus a slight variant of Theorem 7.1 (we consider here $\Omega \subset \mathbb{R}^N$ with $N \geqq 3$, but curl and div are restricted to the first 3 components and the inner product to consider is

$$\sum_{j=1}^{3} \alpha_j \beta_j)$$

makes it possible to pass to the limit in

$$\sum_{j=1}^{3} \frac{\partial y_1}{\partial x_j} d_{1j}(y^{\varepsilon}). \ \square$$

Remark 7.5 Denoting by det M the determinant of the $N \times N$ matrix M, it can be proved that the mapping $y \to \det(\nabla y)$ is continuous from $(W^{1,N}(\Omega))^N$ ($\Omega \subset \mathbb{R}^N$) endowed with its weak topology into $L^1(\Omega)$ endowed with the topology of distributions or the vague topology of measures: see [11], Theorem 4 or [10], Corollary 6.2.2; this follows from the proof given in Remark 7.4 above. In contrast with this assertion is the fact that the mapping $y \to \det(\nabla y)$ is *not* continuous from $(W^{1,N}(\Omega))^N$ endowed with its weak topology into $L^1(\Omega)$ endowed with its *weak* topology (see [8], counterexamples 7.1 and 7.3; the latter is reproduced in example 4.2 above). $\square$

We end this subsection with the example of a weakly lower semicontinuous functional which is of interest in homogenization theory.

Theorem 7.4 Let Ω be an open subset of $\mathbb{R}^N$ and y^{ε} be a sequence such that

$$y^{\varepsilon} \longrightarrow y \text{ in } (W^{1,2}(\Omega))^N \text{ weak.}$$

Then for all $\phi \in \mathcal{D}(\Omega)$, $\phi \geqq 0$:

$$\begin{cases} \liminf_{\varepsilon\to 0} \int_\Omega \phi[\mathrm{tr}({}^t\nabla y^\varepsilon \nabla y^\varepsilon) - (\mathrm{tr}\,\nabla y^\varepsilon)^2]dx \geq \\ \qquad \geq \int_\Omega \phi[\mathrm{tr}({}^t\nabla y\,\nabla y) - (\mathrm{tr}\,\nabla y)^2]dx. \ \square \end{cases} \tag{7.23}$$

This theorem follows from a direct application of Theorem 7.3 (or more exactly of Theorem 3.2, since ∇y^ε is just bounded in $(L^2(\Omega))^{N^2}$): indeed the function f defined by:

$$f(M) = \mathrm{tr}({}^tMM) - (\mathrm{tr}\,M)^2 = \sum_{i,j=1}^N (M_{ij})^2 - (\sum_{i=1}^N M_{ii})^2$$

is quadratic in M and because of Cauchy-Schwarz inequality satisfies:

$$f(a \otimes b) \geq 0, \ \forall a,b \in \mathbb{R}^N.$$

7.3 The case of functions depending on a single variable

This is an example where the constant rank condition (6.2) is not satisfied, but nevertheless the result of Theorem 6.1 is still true.

Let Ω be an open set of $\mathbb{R}^N$ ($N \geq 2$); we consider sequences u^ε such that:

$$u^\varepsilon \longrightarrow u \text{ in } (L^\infty(\Omega))^N \text{ weak } *, \tag{7.24}$$

$$\frac{\partial u_i^\varepsilon}{\partial x_j} \text{ bounded in } L^2(\Omega),\ i,j = 1,\dots,N,\ i \neq j\,. \tag{7.25}$$

The prototype of such sequences consists of functions u_i^ε which only depend on the ith component x_i, i.e.:

$$\begin{aligned} &u_i^\varepsilon(x) = y_i^\varepsilon(x_i)\,, \\ &y_i^\varepsilon \longrightarrow y_i \text{ in } L^\infty(\mathbb{R}) \text{ weak } *\,. \end{aligned} \tag{7.26}$$

The manifold V and cone Λ associated with (7.24), (7.25) are:

$$\begin{aligned} V &= \{(\lambda,\xi) \mid \lambda \in \mathbb{R}^N,\ \xi \in \mathbb{R}^N - \{0\},\ \lambda_i\xi_j = 0, i,j = 1,\dots,N,\ i \neq j\} \\ &= \{(0,\xi) \mid \xi \neq 0\} \cup \bigcup_{i=1}^N \{(\lambda,\xi) \mid \xi_i \neq 0,\ \lambda_i \neq 0,\ \xi_j = \lambda_j = 0 \text{ if } j \neq i\}, \end{aligned}$$

$$\Lambda = \bigcup_{i=1}^N \{\lambda \in \mathbb{R}^N \mid \lambda_j = 0 \text{ if } j \neq i\}. \tag{7.27}$$

So Λ is just the union of the axes, and rank $\{\lambda | (\lambda,\xi) \in V\}$ is either 1 or 0 depending on whether ξ is one of the vectors of the basis or not. Condition (6.2) is violated here.

Theorem 7.5 Let f be a function belonging to $C^0(\mathbb{R}^N;\mathbb{R})$ such that

$$f(u^\varepsilon) \longrightarrow f(u) \text{ in } L^\infty(\Omega) \text{ weak } * \tag{7.28}$$

for all sequences u^ε satisfying (7.24) and (7.25). Then f(a) is a sum with constant coefficients of products of the components of a, in other terms:

$$f(a) = \sum_{\substack{k_1=0 \text{ or } 1\\ k_2=0 \text{ or } 1\\ \vdots\\ k_N=0 \text{ or } 1}} C_{k_1 k_2 \ldots k_N}\, a_1^{k_1} a_2^{k_2} \ldots a_N^{k_N} . \tag{7.29}$$

Conversely, if f has the form (7.29), then (7.28) holds true for all sequences satisfying (7.24) and (7.25). □

Theorem 7.6 Let f be a function belonging to $C^0(\mathbb{R}^N;\mathbb{R})$ such that

$$\begin{cases} \forall \phi \in \mathcal{D}(\Omega),\ \phi \geqq 0, \\ \liminf_{\varepsilon\to 0} \int_\Omega \phi(x) f(u^\varepsilon(x))dx \geqq \int_\Omega \phi(x)\, f(u(x))dx \\ \text{for all sequences } u^\varepsilon \text{ satisfying (7.24) and (7.25).} \end{cases} \tag{7.30}$$

Then f is separately convex, i.e. the function

$$a_i \in \mathbb{R} \to f(a_1,a_2,\ldots,a_N) \in \mathbb{R} \tag{7.31}$$

is convex for every $i = 1,\ldots,N$.

Conversely, if f is separately convex and smooth, then (7.30) holds true.□

In order to prove Theorems 7.5 and 7.6, we need the following:

Lemma 7.1 Let Ω be an open set of $\mathbb{R}^N$ and v^ε and w^ε two sequences such that:

$$\begin{cases} v^\varepsilon \longrightarrow v \text{ in } L^\infty(\Omega) \text{ weak } *, \\ w^\varepsilon \longrightarrow w \text{ in } L^\infty(\Omega) \text{ weak } *, \end{cases} \tag{7.32}$$

$$\begin{cases} \dfrac{\partial v^\varepsilon}{\partial x_k} \text{ bounded in } L^2(\Omega) , \\ \\ \dfrac{\partial w^\varepsilon}{\partial x_j} \text{ bounded in } L^2(\Omega) \; j = 1,2,\dots,N, \; j \neq k , \end{cases} \tag{7.33}$$

where k is some fixed index, $1 \leq k \leq N$. Then

$$v^\varepsilon w^\varepsilon \longrightarrow vw \text{ in } L^\infty(\Omega) \text{ weak } *. \quad \square \tag{7.34}$$

<u>Proof of Lemma 7.1</u>

Let $C = \prod_{i=1}^{N}]c_i,d_i[$ be a cube contained in Ω and y^ε be the function defined on C by

$$y^\varepsilon(x_1,x_2,\dots,x_N) = \int_{c_i}^{x_k} w_\varepsilon(x_1,\dots,x_{i-1},t,x_{i+1},\dots,x_N)dt.$$

Then, on C, $\dfrac{\partial y^\varepsilon}{\partial x_k} = w^\varepsilon$ and $\dfrac{\partial y^\varepsilon}{\partial x_j} = \displaystyle\int_{c_i}^{x_k} \dfrac{\partial w^\varepsilon}{\partial x_j} dt$ for $j \neq k$, which implies that y^ε is bounded in $W^{1,\infty}(C)$, then compact in $L^\infty(C)$. Thus it is easy to pass to the limit in the sense of distributions on C in the following formula:

$$v^\varepsilon w^\varepsilon = v^\varepsilon \frac{\partial y^\varepsilon}{\partial x_k} = \frac{\partial}{\partial x_k}(v^\varepsilon y^\varepsilon) - y^\varepsilon \frac{\partial v^\varepsilon}{\partial x_k} \text{ in } \mathcal{D}'(C).$$

This implies (7.34) in C, thus in Ω. $\square$

<u>Proof of Theorem 7.5</u>

If $u \to f(u)$ is weakly continuous, the "first necessary condition" (2.3) obtained in Theorem 2.2 implies that:

$$t \in \mathbb{R} \to f(a + t\lambda) \text{ is affine } \forall a \in \mathbb{R}^N, \quad \forall \lambda \in \Lambda.$$

Due to the form (7.27) of Λ it follows that:

$$a_i \in \mathbb{R} \to f(a_1,a_2,\dots,a_N) \text{ is affine for each } i = 1,\dots,N,$$

which in turn implies (7.29).

The proof of the converse part is by induction on the degree of the products. Let us for example assume that:

$$u_1^\varepsilon\, u_2^\varepsilon \ldots u_{k-1}^\varepsilon \longrightarrow u_1\, u_2 \ldots u_{k-1} \text{ in } L^\infty(\Omega) \text{ weak } \star \tag{7.35}$$

for some k, $2 \leqq k \leqq N$, and prove that:

$$u_1^\varepsilon\, u_2^\varepsilon \ldots u_{k-1}^\varepsilon\, u_k^\varepsilon \longrightarrow u_1\, u_2 \ldots u_{k-1}\, u_k \text{ in } L^\infty(\Omega) \text{ weak } \star\, . \tag{7.36}$$

But the sequences v^ε and w^ε defined by:

$$v^\varepsilon = u_1^\varepsilon\, u_2^\varepsilon \ldots u_{k-1}^\varepsilon,\ w_\varepsilon = u_k^\varepsilon$$

satisfy the assumptions of Lemma 7.1 (with $w = u_k, v = u_1\, u_2 \ldots u_{k-1}$ by virtue of (7.35)), and (7.36) follows from (7.34). □

Remark 7.6 In this proof we only used the "first necessary condition" of Section 2, and not the more precise condition of Section 5, to obtain the form (7.29) of the functions f generating weakly continuous mappings.

For what concerns the converse part of the proof, it was not possible to use Theorem 6.1 since the constant rank condition (6.2) is violated here. Thus we gave a direct proof which ensures that nevertheless the conclusion of Theorem 6.1 holds true. Note that in the prototype case (7.26) it is very easy to pass to the weak limit in products of functions y_i^ε since each of those just depends on one independent variable. □

Proof of Theorem 7.6

That f is separately convex whenever (7.30) holds true immediately follows from Theorem 2.1 and from the specific form (7.27) of Λ.

Let us prove the converse part, for example, in the case N = 3. Let $a = (a_1,a_2,a_3) \in \mathbb{Q}^3$ be fixed. Using the smoothness and the separately convex character of f, we obtain:

$$\begin{cases} f(u_1^\varepsilon,u_2^\varepsilon,u_3^\varepsilon) \geqq f(a_1,u_2^\varepsilon,u_3^\varepsilon) + \dfrac{\partial f}{\partial u_1}(a_1,u_2^\varepsilon,u_3^\varepsilon)\,(u_1^\varepsilon - a_1) \geqq \\ \qquad \geqq \ldots \geqq f(a_1,a_2,a_3) + \dfrac{\partial f}{\partial u_1}(a_1,u_2^\varepsilon,u_3^\varepsilon)\,(u_1^\varepsilon - a_1) + \\ \qquad + \dfrac{\partial f}{\partial u_2}(a_1,a_2,u_3^\varepsilon)\,(u_2^\varepsilon - a_2) + \dfrac{\partial f}{\partial u_3}(a_1,a_2,a_3)(u_3^\varepsilon - a_3)\ . \end{cases} \tag{7.37}$$

Extracting a diagonal subsequence (note that $a \in \mathbb{Q}^3$) we can still assume that:

$$\begin{cases} f(u^\varepsilon) \rightharpoonup \ell \text{ in } L^\infty(\Omega) \text{ weak } * , \\ v_1^\varepsilon(a) = \dfrac{\partial f}{\partial u_1}(a_1,u_2^\varepsilon,u_3^\varepsilon) \longrightarrow v_1(a) \text{ in } L^\infty(\Omega) \text{ weak } * , \\ v_2(a) = \dfrac{\partial f}{\partial u_2}(a_1,u_2,u_3^\varepsilon) \longrightarrow v_2(a) \text{ in } L^\infty(\Omega) \text{ weak } *. \end{cases}$$

Note that there exists a constant M' independent of a (which depends on the supremum of $|\frac{\partial f}{\partial u_i}|$ on some subset of $\mathbb{R}^3$) such that for every $a \in \mathbb{Q}^3$ with $|a|_{\mathbb{R}^3} \leq M$ (M given) one has:

$$\|v_1(a)\|_{L^\infty(\Omega)} \leq M', \quad \|v_2(a)\|_{L^\infty(\Omega)} \leq M'. \tag{7.38}$$

Using Lemma 7.1 first for the sequences $v_1^\varepsilon(a)$ and $w_1^\varepsilon = u_1^\varepsilon - a_1$, then for the sequences $v_2^\varepsilon(a)$ and $w_2^\varepsilon = u_2^\varepsilon - a_2$, we obtain from (7.37) the following inequality:

$$\ell \geq f(a_1,a_2,a_3) + v_1(a)(u_1-a_1) + v_2(a)(u_2-a_2) + \frac{\partial f}{\partial u_3}(a)(u_3-a_3) \tag{7.39}$$

which holds true for all $a \in \mathbb{Q}^3$ and for almost every $x \in \Omega$.

Consider now a sequence of step functions z^δ such that:

$$\begin{cases} z^\delta(x) \in \mathbb{Q}^3, \ |z^\delta(x)| \leq M, \ \forall x \in \Omega, \quad \forall \delta, \\ z^\delta \longrightarrow u \text{ a.e. in } \Omega. \end{cases}$$

Integrating (7.39) on Ω with respect to $\phi(x)dx$, ($\phi \in \mathcal{D}(\Omega)$, $\phi \geq 0$) yields:

$$\begin{cases} \displaystyle\int_\Omega \phi(x)\ \ell(x)dx \geq \int_\Omega \phi(x)\ f(z^\delta(x))dx\ + \\ \displaystyle\quad + \int_\Omega \phi[v_1(z^\delta)(u_1-z_1^\delta) + v_2(z^\delta)(u_2-z_2^\delta) + \frac{\partial f}{\partial u_3}(z^\delta)(u_3-z_3^\delta)]dx. \end{cases} \tag{7.40}$$

Using Lebesgue's dominated convergence theorem (note that $v_1(z^\delta)$, $v_2(z^\delta)$ and $\frac{\partial f}{\partial u_3}(z^\delta)$ are bounded in $L^\infty(\Omega)$ by virtue of (7.38)), one easily passes to the limit for $\delta \to 0$ in (7.40), which gives (7.30). □

Remark 7.7 Theorem 7.6 is proved in [3] (Proposition 15) in the case N = 2, under the only assumption that f is continuous. □

REFERENCES

[1] Luc Tartar, Une nouvelle méthode de résolution d'équations aux dérivées partielles non linéaires, *Journées d'Analyse Non linéaire, Proceedings, Besançon, France 1977*, ed. by P. Bénilan and J. Robert, Lecture Notes in Mathematics 665, Springer, Berlin (1978), pp. 228-241.

[2] Luc Tartar, Homogénéisation et compacité par compensation, *Séminaire Goulaouic-Schwartz 1978-1979*, Ecole Polytechnique, Palaiseau (1979), Exposé no. 9.

[3] Luc Tartar, Compensated compactness and applications to p.d.e., *Non-linear Analysis and Mechanics, Heriot-Watt Symposium, Vol. IV*, ed. by R.J. Knops, Research Notes in Mathematics 39, Pitman, Boston (1979), pp. 136-212.

[4] Luc Tartar, The compensated compactness method applied to systems of conservation laws, *Systems of Nonlinear Partial Differential Equations*, ed. by J.M. Ball, NATO ASI Series C.111, D. Reidel Publishing Company, Dordrecht (1983), pp. 263-285.

[5] François Murat, Compacité par compensation, Ann. Sc. Norm. Sup. Pisa, 5, (1978), pp. 489-507.

[6] François Murat, Compacité par compensation II, *Proceedings of the International Meeting on Recent Methods in Nonlinear Analysis, Rome, May 8-12 1978*, ed. by E. De Giorgi, E. Magenes and U. Mosco, Pitagora Editrice, Bologna (1979), pp. 245-256.

[7] François Murat, Compacité par compensation: condition nécessaire et suffisante de continuté faible sous une hypothèse de rang constant, Ann. Sc. Norm. Sup. Pisa, 8 (1981), pp. 69-102.

[8] John M. Ball and François Murat, $W^{1,p}$-Quasiconvexity and Variational Problems for Multiple Integrals, J. Funct. Anal., 58, (1984), pp. 225-253.

[9] Charles B. Morrey Jr., Quasiconvexity and the lower semicontinuity of multiple integrals, Pacific J. Math., 2, (1952), pp. 25-53.

[10] John M. Ball, Convexity conditions and existence theorems in non-linear elasticity, Arch. Rat. Mech. Analysis, 63, (1977), pp. 337-403.

[11] Y.G.Reshetnyak, On the stability of conformal mappings in multi-dimensional spaces, Siberian Math. J., 8, (1967), pp. 69-85.

F. Murat
Laboratoire d'Analyse Numérique
Université de Paris VI
75320 Paris Cedex 05
France.

C OLECH

Some remarks concerning controllability

The mathematical theory of optimal control that was born in the late fifties could be briefly described as a refinement of calculus of variations. To the latter, Leonida Tonelli contributed so much. Mathematically, the difference lies in the regularity of the problem. The general formulation of the optimal control problem, introduced by Pontriagin, led also to methods and ideas, which are new even when applied and specified to classical questions of calculus of variations. As an example we could take the notion of controllability and its relation to the necessary conditions of optimality. The necessary conditions in calculus of variations are derived from the principle that the differential or the variation has to be equal to zero at the extreme value of the functional, while in the control problem one considers more explicitly the linear approximation of the problem. The celebrated Pontriagin maximum principle could be expressed shortly by saying that a necessary condition for the optimal solution of a nonlinear problem is the necessary condition for an analogus but linear problem which constitutes a linear approximation of the original one. This is one of the reasons that in optimal control theory the linear problems are more often presented in contrast to calculus of variations. Indeed, the problem of minimizing the functional

$$\int_0^1 (a(t)x(t) + b(t,\dot{x}(t))dt,$$

which is a linear version of the classical variational problem, admits a richer analysis. This analysis was completed rather recently among others, by Lamberto Cesari.

The maximum principle is intimately related to the statement that a local controllability of a solution to a nonlinear control system is implied by the same property of the linear approximation. The latter statement was recently extended by Halina Frankowska to the case of the differential inclusion

$$\dot{x} \in F(x),$$

which is a version of the control system. In this case we need to approximate a set-valued map, and the approximating system she introduced is a differential inclusion

$$\dot{x} \in A(x), \tag{1}$$

where the graph of $A(x)$ is a closed convex cone. Such a set-valued map is called a convex process.

In this lecture we wish to discuss the controllability of system (1); that is, the question when the set reachable from 0,

$$R = \bigcup_{T>0} R_T, \quad R_T = \{x(T) \mid \dot{x}(t) \in A(x(t)),\ x(0) = 0\}$$

is the whole space $\mathbb{R}^n$.

In fact, we are going to present several characterizations of this property obtained recently by J-P. Aubin, H. Frankowska and the author. The proofs and other details are to be found in [1].

Before stating the result we note that system (1) is an extension of the linear control system

$$\dot{x} = Fx + u,\ u \in L, \tag{2}$$

where F is a linear operator from $\mathbb{R}^n$ to itself, and L is a closed convex cone.

Indeed, system (2) is equivalent to the differential inclusion (1) with $A(x) = Fx + L$, and the graph of such A is a convex closed cone. Notice also that, if L is the smallest closed convex cone containing U and such that $0 \in \operatorname{clco} U$, then the controllability of (2) is equivalent to the local controllability (the set reachable from zero contains a neighbourhood of zero) of

$$\dot{x} = Fx + u,\ u \in U. \tag{2'}$$

Thus, the classical case when U is the image of the cube $|v_i| \leq 1$, $i=1,\dots,m$, through a linear map, or where 0 belongs to the interior of $\operatorname{co} U$, corresponds to the case when L in (2) is a subspace. When L is a proper cone, then we are faced with a linear system with "positive controls". Controllability in the first case is characterized by the Kalman controllability condition, which, in our notation, is

$$L + FL + \ldots + F^{n-1}L = \mathbb{R}^n. \qquad (3)$$

Controllability in the second case, first studied by J. Yorke is also characterized in the literature. Condition (3) is then replaced by

$$\dim(L + FL + \ldots + F^{m-1}L) = n \text{ for some } m, \qquad (3')$$

and this is only a necessary condition. The controllability of (2), or equivalently local controllability of (2'), depends on the behaviour of solution of the adjoint linear system

$$-\dot{q}(t) = F^{*}q(t),\ g \in L^{+}. \qquad (4)$$

where

$$L^{+} = \{p \in \mathbb{R}^n \mid \langle p,x\rangle \geqq 0\ \forall x \in L\}$$

is the positive polar of L and F^{*} is the transpose linear map of F.

System (4) is the adjoint system to the differential inclusion (1) with $A(x)= Fx + L$. If A is an arbitrary convex process, then we define the transpose A* by

$$p \in A^{*}(q) \Longleftrightarrow (q,-p) \in (\text{Graph } A)^{+}$$

and the adjoint inclusion by

$$\dot{q}(t) \in -A^{*}(q(t)). \qquad (5)$$

We notice that $A^{*}(q) \neq \emptyset$ if $q \in A(0)^{+}$, so that $A^{*}(q) \neq \emptyset$ for each q only if $A(0) = \{0\}$, and this is the case when A reduces to a linear operator.

Condition (3') with respect to (5) is expressed by

$$\dim A^{m}(0) = n \text{ for some } m \geqq 0, \qquad (6)$$

where $A^{m}(x) = \bigcup_{y \in A^{m-1}(x)} A(y)$ and $A^{1}(x) = A(x)$.

Clearly, $A^{m}(0)$ is a cone and $\dim A^{m}(0)$ is the dimension of the subspace spanned by $A^{m}(0)$. Condition (6) in analogy to (3) is called the rank condition.

A subspace K is invariant with respect to (1) if and only if for each $x_0 \in K$ any solution of (1) starting from x_0 for $t = 0$, is contained in K for $t > 0$.

In analogy to linear operators, we say that a real number is an eigenvalue of a convex process if the image of $(A-\lambda I)$ is different from $\mathbb{R}^n$. Similarly, q is an eigenvector of A if there is λ real such that $\lambda q \in A(q)$

With this notation and introductory remarks we can state now a characterization for the controllability of (1).

<u>Theorem</u> Assume that A(x) in (1) is not empty for each x and that the graph of A is a closed convex cone. Then, the following conditions are equivalent:

(i) the differential inclusion (1) is controllable (i.e., $R = \mathbb{R}^n$),

(ii) for some $T > 0$, $R_T = \mathbb{R}^n$,

(iii) $q(t) \equiv 0$ is the only solution of (5) defined on $[0,T]$ for some $T > 0$,

(iv) A has neither proper invariant subspaces nor eigenvalues,

(v) A* has neither proper invariant subspaces nor eigenvectors,

(vi) the rank condition (6) holds true and A has no eigenvalues,

(vii) the rank condition is satisfied and A* has no eigenvectors,

(viii) for some $m \geq 0$, $A^m(0) = (-A)^m(0) = \mathbb{R}^n$.

The reachable set R_T for (1) when A is a convex process is a convex cone and it increases with respect to T; that is $R_T \subset R_{T'}$, if $T < T'$. The dimension R_T is continuous to the right, nondecreasing, integer-valued and bounded by n. Hence dim R_T = const. for some $[T_0, +\infty)$. This is responsible for the equivalence of (i) and (ii). The equivalence of remaining conditions is obtained from the following duality theorem, which is interesting by itself and rather easy to obtain in the linear case.

Denote by Q_T the set of all q such that q(t) defined on $[0,T]$ is a solution of the adjoint differential inclusion (5) and such that $q(T) = q$. It is rather obvious that Q_T is a closed convex cone.

<u>"Duality" Theorem</u> If A in (1) is a closed convex process with $A(x) \neq \emptyset$ for each x, then Q_T is equal to the positive polar cone of the reachable set R_T of (1).

In fact, the equivalence of (iii) and (ii) follows directly from the above result. To prove the remaining part, it is useful to notice that the closure of the reachable set R is the smallest closed convex cone which is invariant for A, and contains A(0) and R - R, is the smallest invariant subspace containing A(0). Conditions (iv) and (v) as well as (vi) and (vii) are in duality. This follows from the fact that A* has an eigenvector if and only if A has an eigenvalue.

The above result contains as a special case the known characterization for the local controllability of the linear system (2') in the case when $0 \in$ clco U. For details we refer to [1].

REFERENCE

[1] J.-P. Aubin, H. Frankowska, C. Olech, Controllability of convex processes. To appear in SIAM J. on Control and Optimization.

C. Olech
Institute of Mathematics
Polish Academy of Sciences
Warsaw
Poland

J SMOLLER

Symmetry breaking

In this note, we shall summarize the results obtained in papers [1-4].

Our set-up is as follows. We consider the equation

$$\Delta u(x) + f(u(x)) = 0, \quad x \in D_R^n \tag{1}$$

with either homogeneous Dirichlet,

$$u(x) = 0, \ x \in \partial D_R^n, \tag{2}$$

or Neumann

$$\partial u(x)/\partial n = 0, \ x \in \partial D_R^n, \tag{3}$$

boundary conditions. Here D_R^n is an n-ball of radius R. Both of these problems admit radial solutions; i.e. solutions which depend only on the radius $r = |x|$, and thus satisfy the o.d.e.

$$u'' + \frac{n-1}{r} u' + f(u) = 0, \ 0 < r < R, \tag{4}$$

together with the boundary conditions

$$u'(0) = u(R) = 0 \quad \text{(Dirichlet)}, \tag{5}$$

or

$$u'(0) = u'(R) = 0 \quad \text{(Neumann)}. \tag{6}$$

We say that the *symmetry breaks* on the radial solution $\bar{u}(r)$ if this solution bifurcates into an asymmetric solution.

We consider first the case of positive solutions to the Dirichlet problem. From a result in [5], it is known that all such solutions must be radial. Now write (4) as

$$u' = v, \ v' = -(n-1)vr^{-1} - f(u), \tag{7}$$

and denote the solution of (7) with $u(0) = p$, $u'(0) = 0$, by $u(r,p)$. We shall take p as bifurcation parameter, and thus allow the radii R to vary. We set

$$T(p) = \min\ (r > 0 : u(r,p) = 0);$$

then $u(\cdot,p)$ solves (4) and (5) if and only if $T(p) = R$. Next, we need a definition.

Definition $u(\cdot,p)$ is *non-degenerate* if the conditions

$$\Delta w + f'(u(\cdot,p))w = 0 \quad (r < T(p)), \tag{8}$$

and $w = 0$ on $r = T(p)$, imply $w \equiv 0$.

Thus a solution is non-degenerate if and only if 0 is not in the spectrum of the linearized equations. It follows easily from this that bifurcation is possible only on degenerate solutions.

Here is our first theorem, which characterizes positive non-degenerate solutions of the Dirichlet problem.

Theorem 1 ([1]) $u(\cdot,p)$ is non-degenerate if and only if both of the following conditions hold:

(a) $T\ (p) \neq 0$, and (b) $u'(T(p),p) \neq 0$.

It is useful to sketch quickly the ideas in the proof. Thus, suppose (a) and (b) hold, and w solves (8) and $w = 0$ on $r = T(p)$; we shall show that $w \equiv 0$. Now w has a spherical harmonic representation

$$w(r,\theta) = \sum_{N=0}^{\infty} a_n(r)\Phi_N(\theta),\ \theta \in S^{n-1},\ 0 \leq r \leq T(p),$$

where $\tilde{\Delta}\Phi_N = \lambda_N\ \Phi_N$, $\lambda_N = -N(N + n-2)$, and $\tilde{\Delta}$ is the Laplace-Beltrami operator on S^{n-1}. The a_N's satisfy the equations

$$a_N'' + \frac{n-1}{r}\, a_N' + (f'(u) + \frac{\lambda_N}{r^2})a_N = 0,\ 0 < r < T(p),$$

and if $N \geq 1$, $a_N(0) = 0 = a_N(T(p))$. If then, $N > 1$ and $\tilde{R} \leq T(p)$ is the first positive zero of a_N, we have

$$(r^{n-1} a_N')' + (f' + \frac{\lambda_N}{r^2}) r^{n-1} a_N = 0,$$

and

$$(r^{n-1}v')' + (f' + \frac{\lambda_1}{r^2})r^{n-1}v = 0.$$

Subtract these and integrate over $0 \leqq r \leqq \tilde{R}$ to get

$$\tilde{R}^{n-1}a_N'(\tilde{R})\ v\ (\tilde{R}) = \int_0^{\tilde{R}} [-\lambda_N + \lambda_1]r^{n-3}a_N(r)v(r)dr,$$

and if $a_N > 0$ on $0 < r < \tilde{R}$, we see that the left side is positive, while the right side is negative. Thus $a_N \equiv 0$ if $N > 1$. If $N = 1$, we take $\tilde{R} = R$ in the above argument, to get

$$\tilde{R}^{n-1}\ a_1'(R)v(R) = 0$$

and as $v(R) \neq 0$, we have $a_1'(R) = 0$ so $a_1 \equiv 0$. Similarly, one can show that $T'(p) \neq 0$ implies $a_0 \equiv 0$. The converse is proved similarly.

<u>Corollary 2</u> ([1]). The kernel of the linearized operator about a degenerate positive solution of the Dirichlet problem is always of the form $w = a_0(r) + a_1(r)\Phi_1(\theta)$ and $a_0 \equiv 0$ iff $T'(p) = 0$, while $a_1 \equiv 0$ iff $u'(T(p),p) = 0$.

<u>Corollary 3</u> ([1]). A necessary condition for symmetry-breaking on a positive radial solution of the Dirichlet problem is that $f(0) < 0$.

<u>Proof</u> It is easy to see $v'(R) \geqq 0$, where $R = T(p)$. Thus from the equation $v'(r) = -(n-1)r^{-1}v(r) - f(u(r))$, we have, at $r = R$, $0 \leqq -f(u(R)) = -f(0)$. But if $f(0) = 0$, the line $u = 0$, $v = 0$, $r \geqq 0$ is invariant for (7), so no solution enters $u = v = 0$ in finite r. Hence $f(0) = 0$ and the result follows.

There is a problem with proving the existence of positive solutions when $f(0) < 0$; all old proofs fail. We have, however the following theorem.

<u>Theorem 4</u> ([4]). Let $0 < c \leqq f(u)/u^k \leqq d$ as $u \to \infty$, where $0 < k < n/(n-2)$. Then there is a $\bar{p} > 0$ such that if $u(0) \geqq \bar{p}$, positive solutions exist.

This theorem is used to prove the existence of degenerate solutions; see [1] for details.

In order to prove that the symmetry actually breaks, we need a theorem in [6]. This is applied to various functions f where we show that an n-dimensional cone of asymmetric solutions bifurcates out; see [6].

Next, we turn to Neumann boundary conditions. We consider first, the simplest radial solutions; the monotone ones. Let

$$T_N(p) = \inf\,(r > 0 : u'(r,p) = 0),$$

and let $u(\cdot,p)$ be a solution. As before, any element in the kernel of the linearized equations has a spherical harmonic decomposition

$$w = \sum_{n=0}^{\infty} a_n(r)\, \Phi_n(\theta).$$

That there is a big difference between Neumann and Dirichlet boundary conditions is apparent from the next lemma.

<u>Lemma 5</u> ([3]). $a_1 \equiv 0$.

<u>Proof</u> It is easy to see that $a_1 = cu'$, where c is some constant. If $R = T_N(p)$, then $a_1'(R) = cu''(R)$, and if $a_1'(R) = 0$, then $a_1 \equiv 0$. Thus we show $u''(R) \neq 0$. But this is easy since otherwise the equation gives $f(u(R)) = 0$, and we arrive at a contradiction as above.

We also have the following proposition, whose proof we omit.

<u>Proposition 6</u> ([3]): $w = a_0(r) + a_N(r)\Phi_N(\theta)$, where $N > 1$, i.e., at most one asymmetric mode is non-zero.

Note, the eigenspace corresponding to the eigenvalue λ_K has dimension $\ell_N = \binom{N+n-2}{N} \frac{2N+n-2}{N+n-2}$, and hence they grow like binomial coefficients with N. One can pose the question, which mode is non-zero?; i.e. for which N's is $a_N \not\equiv 0$? (Recall for the earlier case, we always have $N = 1$.) That there is no answer to this question follows from an example constructed in [3], which we now describe.

Let $f(u) = u(1-u)$. Then it is not hard to show dom $(T_N) = (0,1)$. We find an integer $k_0 > 0$ such that

(i) $\exists\ (p_k : k \geq k_0) \subset (0,1),\ p_k \to 1$

(ii) If $u_k(r) \equiv u(r,p_k)$, $k \geq k_0$, then

$\exists$ a non-trivial solution to the problem

$$a_k'' + \frac{(n-1)}{r} a_k' + (f'(u_k) + \frac{\lambda_k}{r^2}) a_k = 0, \ 0 < r < T_N(p_k)$$

$$a_k'(0) = a_K'(T_N(p_k) = 0;$$

(iii) On each u_k, the symmetry breaks.

We can only show that the dimensions of the cones of asymmetric bifurcating solutions are n. But we believe that they are readily ℓ_k-dimensional.

ACKNOWLEDGEMENT

The research leading to this paper was supported in part by NSF under Contract No. DMSH3-01243.

REFERENCES

[1] Smoller, J., and A. Wasserman: Existence, uniqueness, and nondegeneracy of positive solutions of semilinear elliptic equations, Comm. Math. Phys. 95, (1984), 129-159.

[2] Smoller, J., and A. Wasserman: Symmetry-breaking for positive solutions of semilinear elliptic equations, Arch. Rat. Mech. Anal., (to appear).

[3] Smoller, J., and A. Wasserman: Symmetry-breaking for solutions of semilinear elliptic equations with general boundary conditions, Comm. Math. Phys., (to appear).

[4] Smoller, J., and A. Wasserman: Some new existence theorems for positive solutions of semilinear elliptic equations on balls, (in preparation).

[5] Gidas, B., Ni, W. and L. Nirenberg: Symmetry of positive solutions of nonlinear elliptic equations in $\mathbb{R}^n$, Comm. Math. Phys. 68, (1979), 202-243.

[6] Vanderbauwhede, A., Local Bifurcation and Symmetry, Res. Notes in Math., No. 75, Boston: Pitman, 1982.

J. Smoller
Department of Mathematics
University of Michigan
Ann Arbor
Michigan 48109
USA

R TEMAM

Variational problems in solid mechanics

INTRODUCTION

The aim of this article is to describe some results concerning a variational inequality of solid mechanics corresponding to the Prandtl-Reuss law of perfect plasticity.

For time independent problems - or what is called in mechanics the deformation theory - the perfectly plastic behaviour of a solid is governed by the variational principle of Hencky [5]. This variational problem and other related models for plates or bars lead to nonstandard problems in the calculus of variations: because of the possible appearance of surface discontinuities for the fields of displacements, these functions are expected to be piecewise smooth and the solutions to the variational problems are therefore sought in spaces of functions whose derivatives are bounded measures. Typically we meet here the space $BV(\Omega)$ of functions with bounded variation (cf. L. Cesari [1]) and similar spaces like $BD(\Omega)$ the space of vector functions with bounded deformation (cf. G. Strang, R. Temam [9]) or the space $HB(\Omega)$ of functions with a bounded hessien [2]. The setting of the Hencky law in these function spaces, the existence of solutions, the derivation of the Euler equations of the problem (i.e., the necessary and sufficient conditions of optimality) are not easy. These questions which have been the object of the efforts of several mathematicians during these last years, are now fairly well understood; the reader is referred to R. Temam [13] for a presentation of the results and bibliographical references.

After a brief survey of the main results and necessary notations for the static case, this paper will be devoted to the presentation of the recent results of the author [14], [15] concerning the time dependent case, i.e., the Prandtl-Reuss law in the quasi-static case. Strictly speaking, the governing equations do not lead in that case to calculus of variations problems; they rather lead to variational inequalities, but the solution of these variational inequalities rely totally on the methods of the calculus of variations and on the methods and results derived for the static case.

Let Ω be an open set of $\mathbb{R}^3$ representing the shape of the solid which undergoes a perfectly plastic deformation. The unknowns of the problem are the field of displacements u and the field of stresses σ (which are respectively a vector and a tensor field on Ω, see below for more details). In the static case u and σ are solutions to two dual variational problems. In the evolution case in which we are more particularly interested here, u and σ depend also on t, the time; $\sigma = \sigma(t)$ is a solution to an evolution variational inequality studied by C. Johnson [6] and P. Suquet [11] while the determination of u depends on coupled equations for u and σ which include an inequality known in mechanics as the principle of maximum dissipation. The results that we present here contain a regularity result for σ which improves the results of [6], [11] and the study of the maximum dissipation principle.

The article contains two parts: Section 1 dealing with the static case and Section 2 corresponding to the time dependent case. The results presented here will be proved in detail in [15].

PLAN

1. THE DEFORMATION THEORY PROBLEM

1.1 Statement of the problem

A solid body occupies at rest a region Ω of $\mathbb{R}^3$ with boundary Γ. This solid is deformed under the action of volumic forces of density f inside Ω and surface forces of surface density F on some part Γ_1 of Γ. The state of the deformed solid is described by the vector field $u = u(x)$ and the tensor field $\sigma = \sigma(x)$, $x = (x_1,x_2,x_3) \in \Omega$; u(x) is the displacement of the point x between the rest and the new equilibrium, $\sigma(x)$ is the Cauchy stress tensor at x.

Under the assumption of small deformations, σ satisfies the equilibrium equation

$$\operatorname{div} \sigma + f = 0 \text{ in } \Omega \tag{1.1}$$

and the boundary condition

$$\sigma \cdot \nu = F \text{ on } \Gamma_1, \tag{1.2}$$

where $\nu = (\nu_1, \nu_2, \nu_3)$ is the unit outward normal on Γ and $\operatorname{div} \sigma$ and $\sigma \cdot \nu$ are the vectors with components

$$(\operatorname{div} \sigma)_i = \sum_{j=1}^{3} \frac{\partial \sigma_{ij}}{\partial x_j}, \quad (\sigma \cdot \nu)_i = \sum_{j=1}^{3} \sigma_{ij} \nu_j$$

The field of displacements u is given on the complement Γ_0 of Γ_1 on Γ:

$$u = U, \; x \in \Gamma_0. \tag{1.3}$$

Besides the basic conditions (1.1) - (1.3), σ and u satisfy some other relations determined by the *constitutive law* which we now state. Let E denote the space of symmetric tensors of order 2; then E is the sum $E^D \oplus RI$, where I is the identity tensor and E^D the subspace of E consisting of tensors with a vanishing trace. We denote by $\xi \cdot \eta$ and $|\xi|$ the scalar product and the norm in E. We are given a convex set K in E. Here, for the sake of simplicity only, we will take the convex K to be

$$K = \{\sigma \in E, \; \sum_{i,j=1}^{3} |\sigma_{ij}^D|^2 \leq 1\}, \tag{1.4}$$

where $\sigma^D = \sigma - \frac{1}{3} I \operatorname{tr}$ is the deviatoric of σ. One of the relations satisfied by σ is

$$\sigma(x) \in K, \; \forall x \in \Omega . \tag{1.5}$$

The other one is a variational inequality connecting σ and the field of strain $\varepsilon(u)$, i.e. the tensor field with components:

$$\varepsilon_{ij}(u) = \frac{1}{2} \left(\frac{\partial u_i}{\partial x_j} + \frac{\partial u_j}{\partial x_i}\right).$$

This relation, slightly simplified for convenience, reads

$$(\varepsilon(u)(x) - \sigma(x)) \cdot (\tau - \sigma(x)) \leq 0, \quad \forall \tau \in K, \quad \forall x \in \Omega. \tag{1.6}$$

It can be shown that u and σ satisfy this complicated set of relations if and only if they are solutions to two separate variational problems which are dual to each other. These problems are recalled in Section 1.2 where we will also briefly recall the main results for these problems.

1.2 The function spaces. The main results

We denote by $L^2(\Omega)$ the space of square integrable real functions on Ω and by $H^1(\Omega)$ the Sobolev space of order one on Ω,

$$H^1(\Omega) = \{u \in L^2(\Omega), \frac{\partial u}{\partial x_i} \in L^2(\Omega), i = 1,2,3\}.$$

We then introduce the two following spaces (cf. (1.1) - (1.3)):

$$C_{ad} = C_{ad}(U) = \{u \in H^1(\Omega)^3, u = U \text{ on } \Gamma_0\}$$

$$S_{ad} = S_{ad}(f,F) = \{\sigma \in L^2(\Omega,E), \text{div } \sigma + f = 0 \text{ in } \Omega, \sigma\cdot\nu = F \text{ on } \Gamma_1\}.$$

We assume that U is given in $H^1(\Omega)^3$; $L^2(\Omega,E)$ is the space of square integrable functions from Ω into E. We recall that if $u \in H^1(\Omega)$, then its trace on Γ denoted $\gamma_0 u$ is defined: $\gamma_0 u$ is square integrable on Γ and $\gamma_0 H^1(\Omega)$ is the space $H^{1/2}(\Gamma)$.

The variational problem determining u is:

(P) To minimize, for u in C_{ad},

$$\int_\Omega \phi(|\varepsilon^D(u)|)dx + \frac{1}{2}\int_\Omega (\text{div } u)^2 dx - L(u) \tag{1.7}$$

where

$$L(u) = \int_\Omega fudx + \int_{\Gamma_1} Fud\Gamma \tag{1.8}$$

and $\psi:\mathbb{R} \to \mathbb{R}$ is the function equal to $\frac{s^2}{2}$ for $|s| \leqq 1$ and to $|s| - \frac{1}{2}$ for $|s| \geqq 1$.

The variational problem determining σ is

(P*) To maximize, for σ in $S_{ad} \cap K_{ad}$,

$$- \frac{1}{2} \int_{\Omega} |\sigma|^2 dx + \int_{\sigma_0} (\sigma.\nu) U \, d\Gamma \tag{1.9}$$

where

$$K_{ad} = \{\sigma \in L^2(\Omega, E), \ \sigma(x) \in K \text{ for a.e. } x \in \Omega\}.$$

The duality between P and P*, the existence of solutions for P* were studied in [10]. The problem P is not expected in general to possess a solution in C_{ad} and for that purpose the space $BD(\Omega)$ was introduced: this is the space of vector functions $u \in L^1(\Omega)^3$ such that $\varepsilon_{ij}(u)$ is a bounded measure, $\forall i,j$ (see [9]).

A suitable modification of problem P is necessary when we consider functions in $BD(\Omega)$. First, for u in $BD(\Omega)$, $\varepsilon^D(u)$ and $|\varepsilon^D(u)|$ are bounded measures. Using the concept of *convex function* of a measure developed in [3], [4] we can define $\psi(|\varepsilon^D(u)|)$ as a bounded measure in Ω. According to [3] this measure is defined as

$$\psi(|\varepsilon^D(u)|) = \psi \circ h dx + \mu^s \tag{1.10}$$

where we have written

$$|\varepsilon^D(u)| = h dx + \mu^s, \tag{1.11}$$

the Lebesgue decomposition of $|\varepsilon^D(u)|$ ($h \in L^1(\Omega)$). Then in (1.7) we replace

$$\int_{\Omega} \psi(|\varepsilon^D(u)|) dx \quad \text{by} \quad \int_{\Omega} \psi(|\varepsilon^D(u)|).$$

Another modification of P which will not be discussed here (see [10], [13]) led us to partly relax the boundary condition to $u \cdot \nu = U \cdot \nu$ on Γ_0 and introduce the relaxed problem Q:

(Q) To maximize, for u in $\tilde{C}_{ad}$,

$$\int_{\Omega} \psi(|\varepsilon^D(u)|) + \int_{\Gamma_0} |T(u-U)| d\Gamma + \frac{1}{2} \int_{\Omega} (\text{div } u)^2 dx - L(u).$$

Here we have

$$\tilde{C}_{ad} = \{u \in U(\Omega),\ u\cdot\nu = U\cdot\nu \text{ on } \Gamma_0\},$$

$$U(\Omega) = \{u \in BD(\Omega),\ \text{div } u \in L^2(\Omega)\},$$

and for a given vector p defined on Γ, T(p) is the tensor defined on Γ of components

$$T_{ij}(p) = \frac{1}{2}(p_i\nu_j + p_j\nu_i).$$

The problem Q is just an extension of P in the sense that we minimize the *same functional* on a larger set, but it was proved in [12] that

$$\inf P = \inf Q.$$

Also the existence of solutions for Q was proved in [12]. Finally the duality between Q and P* was investigated in [8].

We summarize the results proved in the static case in the following theorem:

Theorem 1.1

(i) Duality

Inf P = Inf Q = Sup P* *and this number is finite* (i.e. $> -\infty$) *if and only if the set of admissible* σ *for* P*, $S_{ad} \cap K_{ad}$ *is not empty in which case* P* *possesses a unique solution* σ.

(ii) Solutions for Q

If the safe load condition is satisfied ($S_{ad} \cap \lambda K_{ad} \neq \emptyset$ *for some* λ, $0 < \lambda < 1$, *which implies* $S_{ad} \cap K_{ad} \neq \emptyset$), *then the relaxed problem* Q *possesses at least one solution in* $U(\Omega)$.

(iii) Generalized duality - Optimality condition

For $u \in U(\Omega)$ and $\sigma \in S_{ad}$, *we can define* $\sigma\cdot\varepsilon(u)$ and $\sigma^D\cdot\varepsilon^D(u)$ *as bounded measures in* Ω. *If* u *is a solution of* Q *and* σ *is the solution of* P*, *then we have the generalized optimality conditions*

$$\sigma^D\cdot\varepsilon^D(u) = \psi(|\varepsilon^D(u)|) + \frac{1}{2}|\sigma^D|^2 \text{ in } \Omega \text{ (equality of measures)} \quad (1.12)$$

$$\text{div } u = \frac{1}{3}\text{tr } \sigma \text{ in } \Omega \quad (1.3)$$

$$T^D(U-u) = (U-u)\cdot(\sigma\cdot\nu) \text{ on } \Gamma_0.$$

Conversely, if $u \in C_{ad}$, $\sigma \in S_{ad} \cap K_{ad}$ *satisfy these relations, then* u *is a solution of* Q *and* σ *is the solution of* P^*.

All these results are proved in [13]. See also [2], [13] for similar results for plates. The problems which are essentially left open are the questions of regularity for the solutions of Q. The (piecewise) regularity of u will be particularly difficult to establish; in particular because several free boundaries are involved: that of the plastic region (i.e., the boundary of the region $|\sigma^D(x)| = 1$), and the possible slip surfaces (discontinuity surfaces) for u; see D. Kinderlehrer [7] for some partial results.

2. THE TIME DEPENDENT PROBLEM

We now consider the time dependent problem. The problem is described in Section 2.1 while Section 2.2 contains our main results and some ideas on the proof.

2.1 The Prandtl-Reuss law (quasi-static evolution).

The situation is similar but now f, F, σ, u, U, depend also on time. The relations satisfied by σ and u are similar to (1.1) - (1.5):

$$\text{div } \sigma + f = 0 \text{ in } \Omega, \tag{2.1}$$

$$\sigma\cdot\nu = F \text{ on } \Gamma_1, \tag{2.2}$$

$$u = U \text{ on } \Gamma_0, \tag{2.3}$$

$$\sigma(x,t) \in K, \quad \forall x \in \Omega, \quad \forall t. \tag{2.4}$$

Introducing the sets $S_{ad}(f(t),F(t))$, $C_{ad}(U(t))$, we express these conditions as:

$$\sigma(t) \in S_{ad}(f(t),F(t)) \cap K_{ad}, \; u(t) \in C_{ad}(U(t)). \tag{2.5}$$

We have written $u(t) = u(\cdot,t)$, $f(t) = f(\cdot,t)$, etc. We will denote the derivative with respect to time by a dot so that $\dot{u} = \frac{\partial u}{\partial t}$, etc.

The constitutive law, i.e., the analog of (1.6) is known as the *maximal dissipation principle* and is expressed by

$$(\varepsilon(\dot{u})-\dot{\sigma}) \cdot (\tau-\sigma) \leq 0 \quad \forall\tau \in K \quad (\forall x \in \Omega, \forall t). \tag{2.6}$$

Furthermore, we assume that the following initial values are given

$$u(x,0) = u_0(x), \; x \in \Omega, \tag{2.7}$$

$$\dot{u}(x,0) = u_1(x), \; x \in \Omega, \tag{2.8}$$

$$\sigma(x,0) = \sigma_0(x), \; x \in \Omega. \tag{2.9}$$

Hypotheses

We make on the data the following assumptions

$$f \text{ and } \dot{f} \text{ belong to } L^\infty(0,T;L^2(\Omega)^3), \tag{2.10}$$

$$F \text{ and } \dot{F} \text{ belong to } L^\infty(0,T;L^2(\Gamma_1)^3), \tag{2.11}$$

$$U, \dot{U} \text{ and } \ddot{U} \text{ belong to } L^\infty(0,T;H^1(\Omega)^3), \tag{2.12}$$

$$\begin{aligned} &u_0, u_1 \in H^1(\Omega)^3, \quad u_0 = U(0) \text{ and } u_1 = \dot{U}(0) \text{ on } \Gamma_1, \\ &\sigma_0 \in L^2(\Omega;E), \; \sigma_0(x) \in K \text{ a.e. } x \in \Omega. \end{aligned} \tag{2.13}$$

We also make the following safe load assumption which guarantees that $K_{ad} \cap S_{ad}(f(t),F(t))$ is not empty, $\forall t$ (compare to Theorem 1.1,(ii)):

$$\begin{aligned} &\text{There exists } \chi \text{ such that } \chi, \dot{\chi} \text{ belong to } L^\infty((0,T) \times \Omega;E) \\ &= L^\infty(0,T;L^\infty(\Omega;E)), \; \chi(\cdot,t) \cap S_{ad}(f(t),F(t)) \; \forall t \text{ and, for some} \\ &\delta > 0, \; \chi(x,t) \in \frac{K}{1+\delta}, \; \forall x \in \Omega, \; \forall t \in (0,T). \end{aligned} \tag{2.13}$$

Under these assumptions, the problem that we want to solve is (2.1) - (2.9).

The variational inequality for σ

It can be shown that if σ, u are solutions of P, then σ is a solution of a variational inequality which plays a role similar to problem P* in the static case, and which suffices to determine σ.

The strong form of the inequality determining σ is

$$(\dot{\sigma},\tau-\sigma) \geqq \int_{\Gamma_0} \dot{U} \otimes \nu \cdot (\tau-\sigma)d\Gamma, \ \forall t, \tag{2.15}$$

for every τ such that

$$\tau, \dot{\tau} \in L^2(0,T;L^2(\Omega;E)), \tag{2.16}$$

$$\tau(t) \in K_{ad} \cap S_{ad}(f(t),F(t)), \ \forall t. \tag{2.17}$$

We denote by $(\cdot,\cdot)$ and $|\cdot|$ the scalar product and the norm in $L^2(\Omega,E)$. The weak form of the inequality which can be deduced from (2.14) (cf. [9]) is

$$\frac{1}{2}|\sigma(t)|^2 \leqq \frac{1}{2}|\sigma_0|^2 + (\sigma(t),\tau(t)) - (\sigma_0,\tau(0)) -$$

$$\int_0^t (\sigma(s),\tau(s))ds + \int_0^t \int_{\Gamma_0} \dot{U}(s) \otimes \nu \cdot (\sigma(s)-\tau(s))d\Gamma \, ds,$$

for $0 < t < T$, for every τ satisfying (2.15), (2.16).

It was proved by C. Johnson [6] and P. Suquet [11] that (2.14) - (2.17) possesses a unique solution satisfying

$$\sigma,\dot{\sigma} \in L^\infty(0,T;L^2(\Omega;E)), \tag{2.19}$$

$$\sigma(t) \in K_{ad} \cap S_{ad}(f(t),F(t)), \ \forall t. \tag{2.20}$$

The same result with (2.19) replaced by

$$\sigma \in L^\infty(0,T;L^2(\Omega;E)) \tag{2.21}$$

was proved in [15] for (2.18) (this slightly improves the uniqueness result and is used in the proof of Theorem 2.1).

2.2 Statement of the main results

We have the following result.

Theorem 2.1

(i) Existence uniqueness for σ

Under the above assumptions there exists a unique solution σ *to* (2.17) *which satisfies* (2.18), (2.19) *and furthermore*

$$\sigma \in L^p(\Omega \times (0,T);E), \ \forall p < \infty . \tag{2.22}$$

(ii) Existence for u and the maximal dissipation property

There exists a solution u *to problem* P *which satisfies*

$$u, \dot{u} \in L^\infty(0,T;BD(\Omega)) \tag{2.23}$$

$$\text{div } u, \text{div } \dot{u} \in L^2(0,T;L^2(\Omega)). \tag{2.24}$$

Furthermore for every τ *satisfying*

$$\tau \in L^\infty(0,T;L^2(\Omega;E)), \quad \text{div } \tau \in L^2(0,T;L^3(\Omega)^3) \tag{2.25}$$

$$\tau^D \in L^\infty(0,T;L^\infty(\Omega;E^D)) \tag{2.26}$$

(and in particular $\tau = \sigma$*)* $\varepsilon(\dot{u})\cdot\tau$ *and* $\varepsilon(\dot{u})^D\cdot\tau^D$ *can be defined as bounded measures belonging to* $L^\infty(0,T;M_1(\Omega))$. *Moreover if* $\tau(t) \in K_{ad}$ $\forall t$, *then the maximal dissipation principle*

$$(\varepsilon(\dot{u})-\dot{\sigma})(\sigma-\tau) \leqq 0 \tag{2.27}$$

is satisfied in the sense of measures on $\Omega \times (0,T)$.

The principle of the proof which is long and technical, consists in approximating the problem P by a family of problems P_α, $\alpha > 0$, which are in some sense more regular, and passing to the limit $\alpha \to 0$. For the mechanics point of view, this amounts to replacing perfectly plastic models by slightly viscoplastic models. For the calculus of variations point of view this amounts to replacing the constraint (2.4) by a penalized form of it (in the sense of R. Courant).

For each α, existence and uniqueness of u_α, σ_α are proved. We then pass to the limit $\alpha \to 0$, and u_α, σ_α are shown to converge to u, σ which are solutions of P. The inequality (2.27) is obtained by passing to the limit in a similar inequality for u_α, σ_α . A totally new aspect in this approach and a particularly difficult point of the proof is to show that

$$\sigma_\alpha \to \sigma \text{ in } L^p((0,T) \times \Omega;E),\ \forall p < \infty . \qquad (2.28)$$

For the obtention of (2.28), the introduction of a particular form of the penalization of (2.4) was necessary.

The proof of Theorem 2.1 will appear in [15].

REFERENCES

[1] L. Cesari, *Surface area*, Princeton University Press, 1956.

[2] F. Demengel, Fonctions à hessien borné, *Ann. Institut Fourier Grenoble*, XXXIV(2) (1984), 155-190 and thesis, Université Paris XI, 1986.

[3] F. Demengel and R. Temam, Convex functions of a measure and applications, *Indiana Univ. Math. Journal*, 33 (1984), 273-309.

[4] F. Demengel and R. Temam, Convex functions of a measure. The unbounded case, Journées Pierre de Fermat,Toulouse 1985, Proceedings to be published by Pitman, 1986.

[5] H. Hencky, *Z. Angew. Math. Phys.*, 4 (1924), 323.

[6] C. Johnson, Existence theorems for plasticity problems, *J. Math. Pures Appl.*, 55 (1976), 431-444.

[7] D. Kinderlehrer, Lecture at INRIA, Paris, May 1985.

[8] R. Kohn and R. Temam, Dual spaces of stresses and strains with applications to Hencky Plasticity, *App. Math. Optim.*, 10 (1983), 1-35.

[9] G. Strang and R. Temam, Existence de solutions relaxées pour les équations de la plasticité: étude d'un espace fonctionnel, *C.R. Acad. Sci. Paris*, 287, série A (1978), 515-518, and Functions of bounded deformations, *Arch. Rational Mech. Anal.*, 75 (1980), 7-21.

[10] G . Strang and R. Temam, Duality and relaxation in the variational problems of plasticity, *J. Mécanique*, 19 (1980), 493-527.

[11] P. Suquet, Sur les équations de la plasticité: existence et régularité des solutions, *J. Mécanique*, 20 (1981), 3-39.

[12] R. Temam, Existence theorems for the variational problems of plasticity, in *Nonlinear Problems of Analysis in Geometry and Mechanics*, M. Atteia, D. Bancel and I. Gumowski, Ed., Pitman, London, 1981.

[13] R. Temam, *Mathematical Problems in Plasticity*, Dunod, Paris, 1984 (in French), and Gauthier-Villars, Paris-New York, 1985 (in English).

[14] R. Temam, Principe de dissipation maximale pour la loi de Prandtl-Reuss en plasticité, *C.R. Acad. Sc. Paris*, 302, Serie I, 1986, p.79-82.

[15] R. Temam, A generalized Norton Hoff model and the Prandtl-Reuss law of Plasticity, *Arch. Rat. Mech. Analysis* 1986 and volume dedicated to J. Serrin on his 60th birthday, to appear.

R. Temam
Laboratoire d'Analyse Numérique
Université de Paris XI
91405 Orsay
France

C VINTI

The integrals of the calculus of variations as Weierstrass–Burkill–Cesari integrals

1. INTRODUCTION

In 1911 Tonelli ([46]) emphasized an interesting connection between the functionals of the calculus of variations as Lebesgue integrals and as Weierstrass integrals. However, the mathematicians of the time did not realize the potential of the Weierstrass formulation in the calculus of variations.

This was due to the following main reasons. First, Tonelli himself influenced the choice by opting for the Lebesgue formulation afterwards. Furthermore, at the time the Weierstrass algorithm was rather difficult to deal with. It must be noted that Tonelli obtained the existence of the Weierstrass integral of the calculus of variations by a direct approach only in 1939.

Fifty more years passed before Baiada and Tripiciano ([4] 1957) returned to work on this subject. In the same period Cesari ([32], [33] 1962) introduced a new process of integration for set functions in an abstract setting, suggested by the Weierstrass integral. Actually, it was a far reaching extension of both Weierstrass and Burkill integrals, now known as the Cesari integral. In the same papers Cesari also studied, by means of his new algorithm, a rather general version of Weierstrass multiple integral of the calculus of variations for the parametric case.

Baiada's and Tripiciano's work, and especially the two papers by Cesari, gave a new impulse to the research on the Weierstrass integral of the calculus of variations. We intend to present here the results achieved by the mathematicians who contributed to the development of this theory. But first a point concerning our terminology. By the term Weierstrass integral of the calculus of variations we actually mean the Weierstrass and the Burkill integrals for the one- and the two-dimensional cases, respectively, and also the Cesari integral for the abstract setting.

The interest of this type of integral lies in the possibility of studying the existence of the minima in classes of not necessarily absolutely continuous varieties. In fact, the Weierstrass integral, unlike the Lebesgue integral, always represents the measure of a geometric quantity connected with the variety (see Menger [41] and Aronszain [1]). This is evident even in the particular case of the length integral over a rectifiable continuous curve: the Weierstrass integral can be used for any representation, while the Lebesgue integral applies only for absolutely continuous representations.

2. ON THE INTEGRATION OF SET FUNCTIONS

2.1 The Weierstrass integral

Let $[a,b] \subset \mathbb{R}$ be a fixed interval. Denote by $\{I\}$ the collection of all closed subintervals of $[a,b]$ and by $\mathcal{D}$ the family of all finite subdivisions of $[a,b]$. For every $D \in \mathcal{D}$ we take $\delta(D) = \max\{|I|, I \in D\}$ where $|I|$ is the length of I.

Let $\psi:\{I\} \to \mathbb{R}$ be a given interval function. If the following limit exists in $\mathbb{R}$

$$\lim_{\delta(D)\to 0} \sum_{I\in D} \psi(I),$$

its value is denoted by $\int_a^b \psi$ and is called the *Weierstrass integral* of ψ over $[a,b]$ ([28]).

An existence theorem for the Weierstrass integral was given by Tonelli [47] in terms of the concept of approximate subadditivity. An approximation theorem was also stated for this integral (Vinti, [48]) in the following terms:

$$\int_a^b \psi = \lim_{h\to 0^+} \int_a^{b-h} \psi([x,x+h])h^{-1}\, dx, \tag{1}$$

where the integral on the right-hand side is a Riemann integral.

The approximation theorems for the variation and the length of a continuous curve $f:[a,b] \to \mathbb{R}$, due to Baiada [3] and Radò [43], follows directly from (1). To see this, it is sufficient to consider the particular interval functions

$$\psi([x',x'']) = |f(x'')-f(x')|,$$

$$\psi([x',x'']) = \{|x''-x'|^2 + [f(x'') - f(x')]^2\}^{1/2},$$

whose Weierstrass integrals represent the variation and the length of f. Indeed (1) becomes

$$V_a^b(f) = \lim_{h\to 0^+} \int_a^{b-h} |f(x+h) - f(x)| h^{-1}\, dx,$$

$$L_a^b(f) = \lim_{h\to 0^+} \int_a^{b-h} \{1 + [(f(x+h) - f(x))h^{-1}]^2\}^{1/2}\, dx,$$

respectively, which can also be written as

$$V_a^b(f) = \lim_{h\to 0^+} V_a^{b-h} f_h, \qquad L_a^b(f) = \lim_{h\to 0^+} L_a^{b-h} f_h,$$

where

$$f_h = h^{-1} \int_0^h f(x+t)dt$$

is the integral mean of f.

We mention here that Brandi [17] extended (1) to $\tilde{R}$, and Ragni [44] proved (1) in terms of Lebesgue integration.

2.2 The Burkill integral

Let R_0 be the rectangle $R_0 = [a,b] \times [c,d] \subset \mathbb{R}^2$ and let $\{R\}$ be the collection of all close subrectangles of R_0 of the type $R = I \times J$.

Let us consider a family $\mathcal{D}$ of finite partitions of R_0, and denote by $\delta(D) = \max\{\|R\|,\ R \in D\}$, where $\|R\|$ is the diameter of R.

Let ψ: $\{R\} \to \mathbb{R}$ be a given rectangle function. If the following limit exists in $\mathbb{R}$

$$\lim_{\delta(D)\to 0} \sum_{R\in D} \psi(R),$$

its value is denoted by $\int_{R_0} \psi$ and is called the *Burkill integral* of ψ over R_0 ([28]).

Remarkable cases of this integral can be obtained by particularizing the family $\mathcal{D}$.

In more details, let us consider the following families of partitions of R_o:

(i) $\mathcal{D} = \mathcal{D}_P$ is the family of all the Cartesian partitions of R_0: that is, every element of $\mathcal{D}_P$ is the product of two finite subdivisions of [a,b] and [c,d]. In this case the Burkill integral is denoted by $(E)\int_{R_0} \psi$ and it is called the *extended Burkill integral* (Burkill [28]).

(ii) $\mathcal{D} = \mathcal{D}_J$ is the family of all the intermediate partitions of R_0; that is, every element of $\mathcal{D}_J$ is obtained from a given Cartesian partition of R_0, replacing every rectangle R of this partition with a Cartesian partition of R, and so on for a finite number of steps.

The corresponding Burkill integral is denoted by $(J)\int_{R_0} \psi$ and it is called the *intermediate Burkill integral* (Vinti [51], [54]).

(iii) $\mathcal{D} = \mathcal{D}_E$ is the family of all the partitions of R_0, without any restriction.

In this case we have the so called *strict Burkill integral* which is denoted by $(S)\int_{R_0} \psi$.

It is obvious that integrability in this strict sense ensures integrability in the intermediate sense, and this one ensures the one in the extended sense. A function ψ may be integrable in the extended sense and not in the intermediate one; but a function ψ integrable in the intermediate sense and not integrable in the strict sense is not known.

The Tonelli existence theorem for the Weierstrass integral can be extended to the strict Burkill integral and to the intermediate one (Vinti [51]).

There is also an approximation theorem valid for every Burkill integral (Vinti [51]), expressed by

$$\int_{R_0} \psi = \lim_{(h,k)\to 0^+} \int_{R_{h,k}} \psi([x,x+h] \times [y,y+k])/hk \; dx\, dy,$$

where the integral on the right-hand side is a Riemann integral and $R_{h,k} = [a,b-h] \times [c,d-h]$.

2.3 The Cesari integral

In [32], [33] Cesari proposed a general theory of integration for set functions which represents a far-reaching extension of both Weierstrass and Burkill integrals.

Let A be a nonempty set, and let $\{I\}$ be a collection of nonempty subsets of A that we shall denote as *intervals*. Let $\mathcal{D}$ be a nonempty family of finite systems

$$D = [I] = [I_1, I_2, \ldots, I_p]$$

of sets $I_i \in \{I\}$, with the condition

$$I_i \cap I_j = \emptyset, \; i \neq j,$$

while, if A is a topological space, this condition is replaced by

$$I_i^o \neq \emptyset, \; i = 1, \ldots, p,$$

$$I_i^o \cap \bar{I}_j = \emptyset, \; i \neq j.$$

Let us consider a *mesh function* $\delta : \mathcal{D} \to \mathbb{R}^+$, that is, a real function defined for all systems $D \in \mathcal{D}$ and satisfying the following hypotheses:

(d_1) $0 < \delta(D) < +\infty$, $D \in \mathcal{D}$;

(d_2) for every $\varepsilon > 0$, there is a $D \in \mathcal{D}$, $\delta(D) < \varepsilon$.

Let

$$\psi : \{I\} \to \mathbb{R}^n, \; \psi = (\psi_1, \psi_2, \ldots, \psi_n),$$

be a real-valued interval vector function.

The following limit, if it exists in $\mathbb{R}^n$,

$$\lim_{\delta(D)\to 0} \sum_{I \in D} \psi(I),$$

is denoted by

$$\int_A \psi = \left(\int_A \psi_i, \int_A \psi_2, \ldots, \int_A \psi_n\right)$$

and is called the *Cesari integral* of ψ over A with respect to the family $\mathcal{D}$ and the mesh function δ (Cesari [32], [33] 1962).

Cesari [32] introduced the concepts of quasiadditivity and quasisubadditivity in order to give sufficient conditions for the existence of the above integral.

2.4 The strong and weak Cesari integral

In [57], [58] Warner extended the definition of the Cesari integral to the case of set functions ψ: $\{I\} \to E$, where E is a Hausdorff locally convex topological vector space.

Moreover he gave a weak and strong version of this kind of integration.

In more detail, the following limit, if it exists as an element of E,

$$\lim_{\delta(D) \to 0} \sum_{I \in D} \psi(I),$$

is denoted by $\int_A \psi$ and is called the *strong Cesari integral* of ψ over A (with respect to $\mathcal{D}$ and δ). Let E* be the dual of E. If for every $e^* \in E^*$, the real function $\langle e^*,\psi\rangle$ is Cesari integrable on A then the linear functional $\int_A \psi : E^* \to \mathbb{R}$ defined by

$$\langle \int_A \psi,\ e^*\rangle = \int_A \langle e^*,\psi\rangle$$

is called the *weak Cesari integral* of ψ *over* A.

Brandi and Salvadori ([19]) successively studied the weak Cesari integral in the case of a Banach space E, in particular, they introduced a natural definition of *weak quasiadditivity* which ensures the existence of the weak integral and is less demanding than quasiadditivity.

Moreover, they proved that the weak Cesari integration is actually a process weaker than the strong Cesari integration.

Remark

The Riemann, Lebesgue-Stieltjes, Hellinger, Henstok, Bochner and Dunford integrals are particular cases of the Cesari integral. This fact was proved by Cesari [32] for the Riemann and the Lebesgue-Stieltjes integrals, and by Warner [58] for the Hellinger integral. For the Henstok integral, it was proved by Pucci [42], who also gave a generalization of this integration process. Brandi and Salvadori [19] showed that the Bochner and Dunford integrals are particular cases of the strong and weak Cesari integrals, respectively.

3. APPLICATIONS TO THE INTEGRALS OF THE CALCULUS OF VARIATIONS IN THE PARAMETRIC CASE

3.1 The Weierstrass integral of the calculus of variations

Given a function $F(x,y;x',y') : A \times \mathbb{R}^2 \to \mathbb{R}$, $A \subset \mathbb{R}^2$, with $F(x,y;0,0) = 0$ and a continuous curve $\mathfrak{C} = \{x = x(t),\ y = y(t),\ a \leq t \leq b\} \subset A$, let us consider the interval function $\psi: \{I\} \to \mathbb{R}$ defined by

$$\psi(I) = F[x(t'),\ y(t'),\ x(t'') - x(t'),\ y(t'') - y(t')],\quad I = [t',t''] \subset [a,b].$$

The Weierstrass integral of the interval function ψ, usually denoted by

$$(W)\int_a^b \psi = (W)\int_a^b F[x,y;x',y'],$$

(originally introduced by Weierstrass) is called the *Weierstrass integral of* F *over the curve* $\mathfrak{C}$.

Tonelli [47], by using his concept of approximate subadditivity, proved the existence of such an integral, when F is continuous and positively homogeneous of degree one in (x',y') and the curve $\mathfrak{C}$ is rectifiable. He then stated a representation theorem in terms of a Lebesgue integral

$$(W)\int_a^b F[x,y;x',y'] = \int_0^{L\mathfrak{C}} F[x(s),y(s);\ x'(s),y'(s)]ds,$$

where s is the arc length parameter.

The following approximation theorem can be derived from the general approximation theorem for the Weierstrass integral (Vinti [48]):

$$(W)\int_a^b F = \lim_{h\to 0^+} \int_a^{b-h} F[x(t),y(t),(x(t+h)-x(t))h^{-1},(y(t+h)-y(t))h^{-1}]dt,$$

and this holds for any representation $x = x(t)$, $y = y(t)$ of the rectifiable curve $\mathfrak{C}$. We note that, in turn, from this statement, when t is the arc length parameter, we can derive Tonelli's representation theorem by dominated convergence.

3.2 The Fubini-Tonelli integral for two curves as a Burkill integral

Consider a function $F(x_1,y_1,x_2,y_2;\ x_1',\ y_1';\ x_2',\ y_2'): A_1 \times A_2 \times \mathbb{R}^n \to \mathbb{R}$, $A_1,A_2 \subset \mathbb{R}^2$, with

$$F[x_1,y_1;x_2,y_2;x_1',y_1';0,0] = 0, \quad F[x_1,y_1;x_2,y_2;0,0;x_2',y_2'] = 0$$

and two continuous curves,

$$\mathfrak{C}_1 = \{x_1 = x_1(u),\ y_1 = y_1(u),\ a_0 \leqq u \leqq b_0\} \subset A_1,$$

$$\mathfrak{C}_2 = \{x_2 = x_2(v),\ y_2 = y_2(v),\ c_0 \leqq v \leqq d_0\} \subset A_2.$$

Let us consider the rectangle function $\psi:\{R\} \to \mathbb{R}$ defined by

$$\psi(R) = F[x_1(a),y_1(a);\ x_2(c),\ y_2(c);\ x_1(b)-x_1(a),\ldots;x_2(d)-x_2(c),\ldots],$$
$$R = [a,b] \times [c,d].$$

The Burkill integral of ψ, which is also denoted by

$$(B)\int_{R_0} \psi = (B)\int_{R_0} F[x_1,y_1;\ x_2,y_2;\ x_1',y_1';\ x_2',y_2'],$$

is the Fubini-Tonelli integral of the calculus of variations (as a Burkill integral) (see also Faedo [37]).

The study of this integral started in 1968 (Vinti [51]). If F is continuous and positively homogeneous of degree one, separately in (x_1', y_1') and in (x_2', y_2'), and if $\mathfrak{C}_1$, $\mathfrak{C}_2$ are rectifiable, then the intermediate Burkill integral exists, $(B\text{-}J)\int_{R_0} F$. Moreover, the approximation theorem

$$(B\text{-}J)\int_{R_0} F = \lim_{(h,k)\to 0^+} \int_{R_{oh,k}} F[x_1(u),\ldots,y_2(v);(x_1(u+h)-x_1(u))/h,$$
$$\ldots,(y_2(v+k)-y_2(v))/k]du\, dv$$

holds. From this statement we can derive, as before, the following representation theorem in terms of a Lebesgue integral

$$(B\text{-}J)\int_{R_0} F = \int_0^{L\mathfrak{C}_1}\int_0^{L\mathfrak{C}_2} F[x_1(s),y_1(s),x_2(\tau),y_2(\tau);x_1'(s),\ldots,y_2'(\tau)]ds\, d\tau.$$

Averna [2] obtained the same results, in Cesari's setting, for the extended Burkill integral.

3.3 The multiple integral of the calculus of variations (as a Cesari integral)

Let $f(p,q)$: $K \times \mathbb{R}^n \to \mathbb{R}^k$, $K \subset \mathbb{R}^m$, be given.

Let us denote by $\{I\}$ a collection of subsets of a topological space A, by $\mathcal{D}$ a given family of finite systems, and by $\delta:\mathcal{D} \to \mathbb{R}^+$ a mesh function. Let $p = p(w)$ be a manifold, that is, a map from A to K, and let $\phi:\{I\} \to \mathbb{R}^n$ be an interval function. Let us consider the interval function

$$\psi(I) = f(p(\tau), \phi(I)), \; \tau \in I.$$

The Cesari integral of ψ over A is denoted by $(C)\int_A \psi = (C)\int_A f(p,\phi)$ and is the *multiple integral of the calculus of variations (as a Cesari integral)*, of the function f over the manifold p (Cesari [32]).

It can be thought of as an abstract formulation and a generalization of the Weierstrass integral over a curve.

If $f(p,q)$ is continuous and positively homogeneous of degree one in q, and if ϕ is quasiadditive, then Cesari proved that the function ψ is also quasiadditive. In other words, the concept of quasiadditivity is preserved under the nonlinear map f. Moreover, the integral $(C)\int_A (p,\phi)$ exists and is independent of the choice of $\tau \in I$.

Cesari [33] also proved that it is possible to associate a vector measure ν with values in $\mathbb{R}^n$ to the function ϕ and a scalar measure μ to $\|\phi\|$, under very natural hypotheses on the function ϕ. The measure ν is an extension of the integral of ϕ to the Borel σ-ring $\mathcal{B}$ of A, and the measure μ is an extension of the integral of $\|\phi\|$ to $\mathcal{B}$; moreover, μ is just the total variation of ν. Both measures are regular, and the following integral representation holds:

$$(C)\int_A f(p,\phi) = \int_A f(p,d\nu/d\mu)d\mu,$$

in terms of a Lebesgue-Stieltjes integral, where $d\nu/d\mu$ is the Radon-Nikodym derivative. In other words, the abstract Weierstrass integral $(C)\int_A f(p,\phi)$ has always a representation in terms of a Lebesgue-Stieltjes integral. We obtain, as a particular case, a representation theorem on a continuous surface of finite area or on a rectifiable curve.

The theory of the multiple integral of the calculus of variations in the sense of Cesari has been extensively studied for vector valued functions in the case where K is a metric space, E and F are Banach spaces and E is uniformly convex. In particular, Warner [57] extended the existence theorem given by Cesari and presented many properties of this generalized integral, but he did not deal with any representation theorem. Brandi and Salvadori [21], [22] obtained an analogue of Cesari's representation theorem in terms of a Bochner integral.

We note that Brandi and Salvadori obtained the extension of Cesari's integral to a vector measure for functions $\psi:\{I\} \to E$ by virtue of the theory they had already developed in [23], and which generalizes Cesari's result to Banach spaces.

Moreover, we also note that, in order to obtain the representation statement, Brandi and Salvadori could not use Cesari's process, because of the lack of an inner product in the Banach space E, but they reached the same conclusion by an interesting connection with convergence theorems for martingales of bounded variation within the theory of stochastic processes.

Up to now, all the results were mentioned in view of achieving a complete independence of the integral from the choice of the points $\tau = \tau_I \in I$, so that they could be applied naturally to the case of integrals of the calculus of variations over continuous and BV curves or surfaces. Recently, Brandi and Salvadori [26] sharpened the theory in order to include the case of not necessarily continuous BV varieties.

To achieve this they substituted a generic set function P(I) for the choice function $p(\tau_I)$. In this frame, they were able to prove the existence of the integral, its semicontinuity, and a Tonelli-type characterization theorem. In the case $P(I) = p(\tau_I)$ their results bind the possible choices to a suitable subfamily. For instance, if p is a BV curve in $\mathbb{R}^n$, it is sufficient to avoid a suitable null set in the choice $\tau_I \in I$.

Brandi and Salvadori considered a set function of the type $\psi_q(I) = f(P(I), \phi(I)\cdot Q(I))$. They obtained, as a particular case, the results on existence, semicontinuity, and representation already proved by Boni [9] for the weighted generalized variation (see [14] and [38]).

4. APPLICATIONS TO THE INTEGRALS OF THE CALCULUS OF VARIATIONS IN THE NON-PARAMETRIC CASE

4.1 The Weierstrass integral of the calculus of variations

This integral was introduced in 1960 (Vinti [49]) following McShane's idea

[40] of associating a suitable parametric integral to the classical non-parametric integral of the calculus of variations. Let us denote by $f(x,y,y'); [a,b] \times \mathbb{R}^2 \to \mathbb{R}$ a given function and by F the associated parametric function, that is,

$$F[x,y;\ x',y'] = x'f(x,y,y'/x'),$$

and we take

$$(W)\int_a^b f = (W)\int_a^b F,$$

as the Weierstrass integral of f. Specifically, for any continuous curve

$$\mathfrak{C} = \{y = y(x),\ a \leq x \leq b\},$$

we consider the interval function $\Phi:\{I\} \to \mathbb{R}$, given by

$$\phi([t',t'']) = F[t',y(t'),t''-t,y(t'') - y(t')]$$

$$= (t''-t)f(t',y(t'),(y(t'') - y(t'))/(t''-t')).$$

We have the following existence and approximation theorems (Vinti [49]):

If f is continuous and F admits a continuous extension to $x' \geq 0$, for every rectifiable continuous curve $\mathfrak{C}$, the integral

$$(W)\int_a^b f(x,y,y')$$

exists and we have

$$(W)\int_a^b f(x,y,y') = \lim_{h\to 0^+} \int_a^{b-h} f(x,y(x),(y(x+h)-y(x))/h)dx.$$

For instance, a class of functions for which the integral exists is the following one:

$$f = \phi(x,y)\ \sqrt[n]{(1+y'^n)} + \psi(x,y)\ |y'| + X(x,y),$$

where ϕ, ψ, X are continuous on $[a,b] \times \mathbb{R}$.

Boni [8] characterized a class of functions f for which the associated function F admits a continuous extension as follows.

Let f be continuous and convex in y'; then, the associated function F admits a continuous extension if and only if the following condition is satisfied: for every compact set $B \subset [a,b] \times \mathbb{R}$, there is a constant H_B such that

$$|f(x,y,y')| \leq H_B(1 + |y'|), \quad (x,y) \in B, \ y' \in \mathbb{R},$$

and the function $f(x,y,y')/(1 + |y'|)$ is continuous in $(x,y) \in B$, uniformly with respect to y'.

The following condition is necessary to the existence of the integral (Vinti [49], [52]).

Assume that

(v_1) f is convex in y'; but is not a straight line in y';

(v_2) for every compact set $B \subset [a,b] \times \mathbb{R}$, the function f is continuous in B, uniformly with respect to y'.

Assume that $\mathfrak{C}$ is a continuous curve and the integral $(W) \int_a^b f(x,y,y')$ exists. Then, $\mathfrak{C}$ is rectifiable.

It is not possible to invert this theorem, so as to have an existence theorem for the integral $(W) \int_a^b f(x,y,y')$ when $\mathfrak{C}$ is continuous and rectifiable. There are, in fact, many examples (Vinti [49]) which show that the integral $(W) \int_a^b f(x,y,y')$, with f satisfying the hypotheses (v_1) and (v_2), does not exist even if $\mathfrak{C}$ is AC.

We proposed to study the problem of the existence of the integral $(W)\int_a^b f$ for the case where the associated function F does not admit a continuous extension.

Boni [7] dealt with this matter in the one-dimensional case, for AC curves and essentially by Cesari's concept of quasiadditivity and quasi-subadditivity. The main result was generalized and extended to the Fubini-Tonelli integral by Boni and Gori, and it will be presented in such formulation in the following section.

Then, Boni [7] studied the existence of the integral $(W) \int_a^b f$ for continuous curves of bounded variation and in the case where the associated function F does not admit of a continuous extension, since otherwise the integral $(W) \int_a^b f$ would exist by virtue of Vinti's existence theorem. The chief results of Boni are the following ones and they are obtained essentially by means of Cesari's concept of quasisubadditivity:

if $f(y')$ is non-negative and convex and if $y = y(x)$, $x \in [a,b]$, is continuous, then the integral $(W)\int_a^b f$ exists (finite or not);

if $f(x,y,y') \geqq 0$ is continuous and convex in y', $x \in [a,b]$, $y,y' \in \mathbb{R}$, and for every compact set $B \subset [a,b] \times \mathbb{R}$, the function $f/(1 + |y'|)$ is continuous in $(x,y) \in B$, uniformly with respect to y', then the integral $(W)\int_a^b f$ exists (finite or not) on every continuous curve of bounded variation.

4.2 The Fubini-Tonelli integral for two curves as a Burkill integral

Let $f(x_1,y_1;x_2,y_2;y_1',y_2') : [a,b] \times [c,d] \times \mathbb{R}^2 \to \mathbb{R}$ be given.

Analogously to the one-dimensional case, we introduce the associated parametric function F, defined, for $x_1' > 0$, $x_2' > 0$, by

$$F[x_1,y_1;x_2,y_2;x_1',y_1';x_2',y_2'] = x_1'\ x_2'\ f(x_1,y_1,x_2,y_2;y_1'/x_1',y_2'/x_2').$$

If $\mathfrak{C}_1 = \{y_1 = y_1(x_1),\ a \leqq x_1 \leqq b\}$,

$\mathfrak{C}_2 = \{y_2 = y_2(x_2),\ c \leqq x_2 \leqq d\}$

are two continuous curves, consider the rectangle function $\phi:\{R\} \to \mathbb{R}$, defined by

$$\phi(R) = F[u',y_1(u'),v',y_2(v');\ u''-u',y_1(u'')-y_1(u'),\ v''-v',\ y_2(v'')-y_2(v')]$$

$$= (u''-u')(v''-v')\ f[u',y_1(u'),v',y_2(v'),(y_1(u'')-y_1(u'))/(u''-u'),\ (y_2(v'')-y_2(v'))/(v''-v')],\quad R = [u',u''] \times [v',v''].$$

The Burkill integral of the function ϕ over R_o is the *Fubini-Tonelli integral of the function* f *over the curves* $(\mathfrak{C}_1,\mathfrak{C}_2)$ *(as a Burkill integral) and is denoted by* $(B)\int_{R_o} f$. The study of this integral started in 1968 (Vinti [52]).

An existence theorem, under the hypothesis that the associated function F admits a continuous extension on the set

$$E = \{P, x_1', y_1', x_2', y_2' : P \in [a,b] \times \mathbb{R} \times [c,d] \times \mathbb{R},$$
$$x_1' \geqq 0,\ x_2' \geqq 0,\ y_1', y_2' \in \mathbb{R}\},$$

analogous to the one-dimensional case, was proved for the integral (B) $\int_{R_0} f$, in the intermediate sense.

The necessary (but not sufficient) condition for the existence of the integral in the one-dimensional case, was extended to the integral (B)$\int_{R_0} f$, in the intermediate and extended cases, under the hypotheses that conditions (v_1) and (v_2) are separately satisfied with respect to y_1' and y_2'.

As in the one-dimensional case, the problem arose studying the existence of the integral (B) $\int_{R_0} f$ when the associated function F does not admit a continuous extension.

Boni and Gori [12], using essentially Cesari's concept of quasiadditivity provided the following wide class of functions for which the integral (B)$\int_{R_0} f$ exists in the intermediate sense, when $\mathfrak{C}_1$ and $\mathfrak{C}_2$ are AC.

Let

$$f(x_1, y_1, x_2, y_2;\ x_1', y_2') \geqq 0$$

be a globally continuous function which is convex with respect to x' and y' separately; let $\mathfrak{C}_1$, $\mathfrak{C}_2$ be two AC curves such that

$$f(x_1, y_1(x_1), x_2, y_2(x_2);\ y_1'(x_1), y_2'(x_2)) \in L_1(R_0).$$

Suppose that a real function

$$\lambda(x_1, y_1, x_2, y_2;\ y_1', y_2') \geqq 0$$

and a number $\sigma > 0$ exist such that

$$\lambda(x_1, y_1(x_1), x_2, y_2(x_2),\ y_1'(x_1), y_2'(x_2)) \in L_1(R_0),$$

and

$$|f(x_1,y_1(x_1),x_2,y_2(x_2);y_1'(\xi),y_2'(\eta))-f(\tilde{x}_1,y_1(\tilde{x}_1),\tilde{x}_2,y_2(\tilde{x}_2);y_1'(\xi),$$
$$y_2'(\eta))| \leqq \lambda(x_1,y_1(x_1),x_2,y_2(x_2);y_1'(\xi),y_2'(\eta)),$$

with

$$|x_1-\tilde{x}_1| < \sigma , \quad |x_2-\tilde{x}_2| < \sigma ,$$

and for every $(\xi,\eta) \in R_o$. Under these hypotheses, the integral $(B)\int_{R_o} f$ exists in the intermediate sense, and the following representation holds:

$$(B) \int_{R_o} f = \int_{R_o} f(x_1,y_1(x_1),x_2,y_2(x_2);\ y_1'(x_1),y_2'(x_2))dx_1\ dx_2.$$

This set of integrands includes, in particular, the class introduced by Boni [7] in the one-dimensional case, a class provided by Breckenridge [27], and a class due to Serrin [45].

By using a weak version of Tonelli's approximate subadditivity, Boni and Ragni [13] found a class of functions for which the existence of the integral $(B) \int_{R_o} f$ is assured in the strict sense (and therefore also in the intermediate and extended senses). They do not use hypotheses of positive homogeneity and convexity on the integrand f, but they work in a class of curves, denoted by AC*, which is more restrictive than the class of AC curves.

Let us consider a continuous function $f(x_1,y_1,x_2,y_2;y_1',y_2')$ and two curves $\mathfrak{C}_1,\mathfrak{C}_2$ in AC*. If

$$\lambda(y_1',y_2') \geqq 0, \quad y_1',y_2' \in \mathbb{R},$$

is convex with respect to y_1' and y_2' separately, if

$$\lambda(y_1'(x_1),\ y_2'(x_2)) \in L_1(R_o),$$

and

$$f(x_1,y_1(x_1),\ldots,y_1',y_2') \leqq \lambda(y_1',y_2'),$$

$$x_1 \in [a,b],\ x_2 \in [c,d],\ y_1',y_2' \in \mathbb{R},$$

then the integral $(B)\int_{R_o} f$ exists in the strict sense and the following representation holds:

$$(B)\int_{R_0} f = \int_{R_0} f(x_1,y_1(x_1),x_2,y_2(x_2);y_1'(x_1),y_2'(x_2))dx_1\, dx_2.$$

Candeloro and Pucci [29] studied the integral $(B)\int_{R_0} f$ in the extended sense, over continuous curves of bounded variation, obtaining some existence and approximation theorems. These results are formally analogous to those proved by the same authors for the multiple integral of the calculus of variations that we present in the following section.

4.3 The multiple integral of the calculus of variations (as a Burkill integral)

Let $f(x,y,u,p,q) : R_0 \times \mathbb{R}^3 \to \mathbb{R}$ be given and let

$$S = \{u = u(x,y),(x,y) \in R_0\},$$

be a continuous surface. Consider the rectangle function $\phi:\{R\} \to \mathbb{R}$ defined by

$$\phi(R) = |R|f(a,c,u(a,c),\gamma_1(R)/|R|,\ \gamma_2(R)/|R|),\ R = [a,b] \times [c,d];$$

where $\gamma_1(R)$ and $\gamma_2(R)$ are the Geöcze expressions

$$\gamma_1(R) = \int_c^d |u(b,y)-u(a,y)|dy,\ \gamma_2(R) = \int_a^b |u(x,d)-u(x,c)|dx.$$

The Burkill integral of the function ϕ over R_0 is the *multiple integral of the calculus of variations (as a Burkill integral)* and it is denoted by $(B)\int_{R_0} f$.

Boni, Gori, and Ragni initiated the study of this integral. Boni and Gori [12] found a class of functions for which the existence of the integral $(B)\int_{R_0} f$ is assured in the intermediate sense when the surface S is absolutely continuous in the sense of Tonelli (ACT). The definition of this class is a simple modification of the one that Boni and Gori introduced for the Fubini-Tonelli integral (see Section 4.2). Boni and Ragni [13] then provided a class of functions f for which the existence of the integral $(B)\int_{R_0} f$ in the strict sense is assured (and therefore also in the intermediate and extended senses) over a surface more regular than ACT. The existence and representation theorems are also a simple modification of the theorems that they stated for the Fubini-Tonelli integrals (see Section 4.2).

Candeloro and Pucci [30] examined the multiple integral of the calculus of variations as a Burkill integral over continuous surfaces of bounded variation in the sense of Tonelli (CBVT). Such integral, under suitable conditions, can be regarded as a Serrin integral [45]. Candeloro and Pucci also proved existence, approximation, and lower semicontinuity theorems for this integral and, in particular, they proved some Tonelli-type theorems. The main statements are the following ones.

Theorem 1 Let $f(x,y,u,p,q)$ be convex in (p,q). Suppose that, for every compact set $B \subset R_0 \times \mathbb{R}$, there exists a constant $K_B > 0$ such that

$$|f(x,y,u,p,q)| \leqq K_B(1 + |p| + |q|), \ (x,y,u) \in B, \quad p,q \in \mathbb{R},$$

and the function $f/(1 + |p| + |q|)$ is continuous on B, uniformly with respect to (p,q). Then, for every surface $u \in$ CBVT, the integral $(B)\int_{R_0} f$ exists in the extended sense, and we have

$$(B)\int_{R_0} f = \lim_{(h,k)\to 0^+} \int_{R_{0_{h,k}}} f(x,y,u_{h,k},u'_{h,k,x},u'_{h,k,y})dx\,dy$$

$$= \lim_{(h,k)\to 0^+} I[u_{h,k}, R_{0_{h,k}}],$$

where

$$u_{h,k} = (hk)^{-1} \int_0^h \int_0^k u(x+\xi,\ y+\eta)\ d\xi\ d\eta$$

is the integral mean of u, defined on

$$R_{0_{h,k}} = [a_0,b_0-h] \times [c_0,d_0-k].$$

Moreover, if

$$f(x,y,u,p,q) \geqq 0,$$

we have

$$(B)\int_{R_0} f \geq \int_{R_0} f(x,y,u,u'_x, u'_y)dx\,dy = I[u,R_0];$$

and for every sequence $(u_n)_{n\in\mathbb{N}}$, $u_n \in$ CBTV, converging in L_1 to u, we also have

$$(B)\int_{R_0} f \leq \liminf_{n\to\infty}(B)\int_{R_0} f_n,$$

where $(B)\int_{R_0} f_n$ denotes the Burkill integral of f over the surface u_n;

Theorem 2 Let f(x,y,u,p,q) be non-negative and convex in (p,q). Suppose that, for every compact set $B \subset R_0 \times \mathbb{R}$, there exist two constants $H_B > 0$ and $K_B > 0$ such that

$$H_B(|p| + |q|) - 1 \leq f(x,y,u,p,q) \leq K_B(1 + |p| + |q|),$$

$$(x,y,u) \in B,\ p,q \in \mathbb{R},$$

and the function f is continuous on B, uniformly with respect to (p,q). Then, for every continuous surface u, the integral $(B)\int_{R_0} f$ exists (finite or not), is the extended sense, and the following approximation statement holds:

$$(B)\int_{R_0} f = \lim_{(h,k)\to 0^+} I[u_{h,k}, R_{0_{h,k}}].$$

Moreover, we have, under these hypotheses, the following properties which contain the classical Tonelli theorems on area:

$$(B)\int_{R_0} f < +\infty \quad \text{if and only if u is CBVT;}$$

$$(B)\int_{R_0} f < +\infty \quad \text{implies } (B)\int_{R_0} f \geq I[u,R_0];$$

$$(B)\int_{R_0} f = I[u,R_0] < +\infty \text{ if and only if u is ACT.}$$

The integral $(B)\int_{R_0} f$ is in the extended sense, while, when f depends only on the pair (p,q), the same integral is in the strict sense.

Theorem 3 If the function $f(p,q) \geq 0$ is convex in (p,q), and there exist two constants $H > 0$ and $K > 0$ such that

$$H(|p| + |q|) - 1 \leq f(p,q) \leq K(1 + |p| + |q|),$$

then the following approximation properties hold for the integral (B) $\int_{R_0} f$ in the strict sense:

(i) If $u \in$ CBVT, we have

$$(B)\int_{R_0} f = \lim_{h\to 0^+} \lim_{k\to 0^+} \int_{R_0} f(u(x+h,y)-u(x,y))h^{-1}, h^{-1}\int_x^{x+h} (u(\xi,y+k)$$

$$- u(\xi,y))k^{-1}\, d\xi)dx\, dy.$$

(ii) If u is absolutely continuous in the sense of Young, we have

$$(B)\int_{R_0} f = \lim_{(h,k)\to 0^+} \int_{R_0} f((u(x+h;y)-u(x,y))h^{-1},(u(x,y+k)-u(x,y))k^{-1})dxdy.$$

Property (i) generalizes a theorem in surface area theory (Vinti [50], [53]); property (ii) extends properties also of surface area theory (Young [59]).

4.4 The multiple integral of the calculus of variations (as a Cesari integral)

Boni and Brandi [10] started to study the multiple integral of the calculus of variations as a Cesari integral, in the non-parametric case, in a very general setting, paying particular attention to existence theorems.

Let A be a set or a topological space, and let $\{I\}$ be a collection of non-empty subsets of A that we shall call *intervals*. Consider a family $\mathcal{D}$ of finite systems $D = [I]$ and a mesh function $\delta:\mathcal{D} \to \mathbb{R}^+$; see the definition of Cesari integral in Section 2.3. Let $\mathcal{A}$ be a σ-ring on A such that $\{I\} \subset \mathcal{A}$. Let us consider a function

$$f(x,p) : K \times \mathbb{R}^n \to R,$$

where K is a compact metric space, $x : \{I\} \to K$ a mapping, $\mu:\mathcal{A} \to \mathbb{R}^n$ a vector measure, and $\nu:\mathcal{A} \to \mathbb{R}_0^+$ a measure. The Cesari integral of the interval function

$$\psi(I) = \nu(I)f(x(I), \mu(I)/\nu(I)),$$

over A, denoted by (C) $\int_A f\,(x,p)$ is the *multiple integral of the calculus of variations as a Cesari integral* in the non-parametric case.

Boni and Brandi [10] prove existence theorems for this integral which extend those given by Boni [7] and by Breckenridge [27] in the one dimensional case. They also extend the statement concerning the existence of the multiple integral of the calculus of variation as a Burkill integral proved by Boni and Gori [12] in the intermediate as well as in the strict case. Finally, the same theorems of Boni and Brandi extend the statement concerning the existence of the multiple integral as a Burkill integral proved by Boni and Ragni [13] for convex integrands and over ACT surfaces without any other regularity assumptions.

Brandi and Salvadori ([37], [38]) further generalized the formulation above. They considered a function $f:K \times E \to \mathbb{R}$ where K is a metric space and E is a Banach space, and two interval functions $\phi:\{I\} \to E_1$ and $\eta:\{I\} \to \mathbb{R}^+$, where E_1 is a Banach space (instead of two measures μ,ν), and a mapping $\tilde{x} : A \times \{I\} \to K$. Then, they introduced a function $g:A \times \{I\} \to E_2$ where E_2 is a Banach space, and studied the Cesari integral, over A, of the interval function

$$\tilde{\psi}(I) = \eta(I)\ f(\tilde{x}(\xi_I,I),\ (\phi(I)/\eta(I)) \cdot g(\xi_I',I)),$$

where ξ_I, $\xi_I' \in I$ and $e_1.e_2:E_1 \times E_2 \to E$ is a bilinear form.

Brandi and Salvadori proved some existence results and also stated a Tonelli-type theorem, and consequently a representation result, in terms of a Lebesgue-Stieltjes integral. This last result is achieved by means of a connection with convergence theorems for martingales, within the theory of stochastic processes.

Concerning the interval functions ϕ and η above, Brandi and Salvadori required a condition which is slightly stronger than quasiadditivity, and which is also preserved, as the quasiadditivity (cf. Section 3.3), by the nonlinear transformation f. Thus, they obtained the existence of the integral by using Cesari's quasiadditivity. Also, Brandi and Salvadori avail themselves of the connection between the non-parametric and the parametric case mentioned in Section 4.1.

The existence theorems which they state contain those given by Boni [8] for continuous curves of bounded variation and the results given by Candeloro and Pucci [30] (see Section 4.3) for the part regarding the existence of the integral which is now considered in the strict sense.

In particular, Brandi and Salvadori ([24],[25]) show that, by their formulation they can establish connections with the integrals of the calculus of variations depending on differential elements of the second order (not of any order $n > 1$ as I wrote in [55], [56] over curves of the space). These classical integrals of the calculus of variations, with differential elements of higher order, were introduced and studied by Cinquini [34], [35], [36]), Berruti Onesti ([5], [6]) and Borgogno ([15], [16]).

Moreover, as a natural application of their integral, Brandi and Salvadori ([38]) introduce the weighted generalized area of a summable surface. This concept extends the generalized area [31] and is inspired by the weighted variation of a curve introduced by Kaiser [39] and then intensively studied by many authors (see [11], [9], [18], [14], [20]).

Recently Brandi and Salvadori [26]) sharpened their results in order to include those integrals of the calculus of variations over curves and surfaces not necessarily continuous.

REFERENCES

[1] Aronszain, N., *Quelques recherches sur l'intégrale de Weierstrass*, C.R. Paris, 205, 1939, 490.

[2] Averna, A., *Teoremi di Esistenza e Approssimazione per l'integrale alla Weierstrass-Burkill-Cesari sopra una Coppia di Varietà*, Atti del Seminario Matematico e Fisico dell'Università di Modena, Vol. 22, pp. 275-334, 1973.

[3] Baiada, E., *La Variazione Totale, la Lunghezza di una Curva, e l'Integrale del Calcolo delle Variazioni*, Rendiconti dell'Accademia Nazionale dei Lincei, Vol. 22, pp. 584-588, 1957.

[4] Baiada, E., and Tripiciano, G., *Un integrale analogo a quello di Weierstrass nel Calcolo delle Variazioni in una variabile*, Rendiconti del Circolo Matematico di Palermo, Vol. 6, 1957, pp. 263-270.

[5] Berruti Onesti, N., *Sopra una classe di Problemi Variazionali di Ordine n*, Annali di Matematica Pura ed Applicata, Vol. 91, pp. 129-161, 1972.

[6] Berruti Onesti, N., *Sopra l'Esistenza dell'Estremo Assoluto per una Classe di Problemi Variazionali di Ordine n*, Rendiconti del Circolo Matematico di Palermo, Vol. 24, pp. 195-228, 1975.

[7] Boni, M., *L'Integrale di Weierstrass Nonparametrico e Quasiadditività*, Rendiconti del Circolo Matematico di Palermo, Vol. 22, pp. 128-144, 1973.

[8] Boni, M., *Quasiadditività e Quasisubadditività nell'Integrale Ordinario del Calcolo delle Variazioni alla Weierstrass*, Rendiconti dell'Istituto Matematico dell'Università di Trieste, Vol. 6, pp. 51-70, 1974.

[9] Boni, M., *Variazione Generalizzata e Quasiadditività*, Atti del Seminario Matematico e Fisico dell'Università di Modena, Vol. 25, pp. 195-210, 1976.

[10] Boni, M., and Brandi, P., *Teoremi di Esistenza per l'Integrale del Calcolo delle Variazioni nel Caso Ordinario*, Atti del Seminario Matematico e Fisico dell'Università di Modena, Vol. 23, pp. 308-327, 1974.

[11] Boni, M., and Brandi, P., *Variazione Classica e Generalizzata con Peso*, Atti del Seminario Matematico e Fisico dell'Università di Modena, Vol. 23, pp. 286-307, 1974.

[12] Boni, M., and Gori, C., *Quasiadditività e Integrali Nonparametrici 2-Dimensionali del Calcolo delle Variazioni*, Rendiconti del Circolo Matematico di Palermo, Vol. 22, pp. 217-238, 1973.

[13] Boni, M., and Ragni, M., *Approssimata Subadditività e Integrali Nonparametrici 2-Dimensionali del Calcolo delle Variazioni*, Rendiconti del Circolo Matematico di Palermo, Vol. 22, pp. 289-312, 1973.

[14] Boni, M., and Salvadori, A., *Convergenza in Lunghezza e Variazione in Senso Generalizzato con Peso*, Atti del Seminario Matematico e Fisico dell'Universita di Modena, Vol. 26, pp. 49-69, 1977.

[15] Borgogno, M., *Sopra la Semicontinuità degli Integrali Curvilinei dei Problemi Variazionali in forma Parametrica di Ordine n*, Annali di Matematica Pura ed Applicata, Vol. 113, pp. 53-97, 1977.

[16] Borgogno, M., *Sopra l'Esistenza dell'Estremo Assoluto per una Classe di Problemi di Lagrange in Forma Parametrica di Ordine Superiore al Secondo*, Atti del Seminario Matematico e Fisico dell'Università di Modena, Vol. 29, pp. 140-165, 1980.

[17] Brandi, P., *Teoremi di Semicontinuità Inferiore e di Approssimazione per le Variazioni con Peso in* $\widetilde{\mathbb{R}}$, Rendiconti del Circolo Matematico di Palermo, Vol. 25, pp. 5-26, 1976.

[18] Brandi, P., *Variazione, Classica e Generalizzata, Rispetto ad una Misura con Derivata Debolmente Misurabile*, Atti del Seminario Matematico e

Fisico dell'Università di Modena, Vol. 28, pp. 63-85, 1979.

[19] Brandi, P., and Salvadori, A., *Sull'Integrale Debole alla Burkill-Cesari*, Atti del Seminario Matematico e Fisico dell'Università di Modena, Vol. 27, pp. 14-38, 1978.

[20] Brandi, P., and Salvadori, A., *Sull'area Generalizzata*, Atti del Seminario Matematico e Fisico dell'Università di Modena, Vol. 28, pp. 33-62, 1979.

[21] Brandi, P., and Salvadori, A., *Martingale ed Integrale alla Burkill-Cesari*, Rendiconti dell'Accademia Nazionale dei Lincei, Vol. 67, pp. 197-203, 1979.

[22] Brandi, P., and Salvadori, A., *Un Teorema di Rappresentazione per l'integrale Parametrico del Calcolo delle Variazioni alla Weierstrass*, Annali di Matematica Pura ed Applicata, Vol. 124, pp. 39-58, 1980.

[23] Brandi, P., and Salvadori, A., *Sull'Estensione dell'Integrale Debole di Burkill-Cesari ad una Misura*, Rendiconti del Circolo Matematico di Palermo, Vol. 30, pp. 207-234, 1981.

[24] Brandi, P., and Salvadori, A., *The Non-parametric Integral of the Calculus of Variations as a Weiestrass Integral*, I, *Existence and Representation*, J. Math. Anal. Appl., Vol. 107, pp. 67-95, 1985.

[25] Brandi, P., and Salvadori, A., *The Nonparametric Integral of the Calculus of Variations as a Weierstrass Integral, II, Some Applications*, J. Math. Anal. Appl., Vol. 112, pp. 290-313, 1985.

[26] Brandi, P., and Salvadori, A., *Existence, semicontinuity and representation for the integrals of the Calculus of Variations. The BV case*, Rend. Circ. Mat. Palermo. Atti Convegno I Centenario, 1984.

[27] Breckenridge, J.C., *Contributions to the Theory of Nonparametric Weierstrass Integrals*, Atti del Seminario Matematico e Fisico dell' Università di Modena, Vol. 21, pp. 145-155, 1972.

[28] Burkill, J.C., *Functions of Intervals*, Proceedings of the London Mathematical Society, Vol. 22, pp. 275-310, 1923.

[29] Candeloro, D., and Pucci, P., *L'integrale di Burkill-Cesari su un Rettangolo e Applicazioni all'Integrale di Fubini-Tonelli Relativamente a Coppie di curve Continue*, Bollettino della Unione Matematica Italiana, Vol. 17B, pp. 835-859, 1980.

[30] Candeloro, D., and Pucci, P., *L'integrale di Burkill-Cesari come Integrale del Calcolo delle Variazioni*, Bollettino della Unione Matematica Italiana, Vol. 18B, pp. 1-24, 1981.

[31] Cesari, L., *Sulle Funzioni a Variazione Limitata*, Annali della Scuola Normale Superiore di Pisa, Vol. 5, pp. 229-313, 1936.

[32] Cesari, L., *Quasiadditive Set Funtions and the Concept of Integral over a Variety*, Transactions of the American Mathematical Society, Vol. 102, pp. 94-113, 1962.

[33] Cesari, L., *Extension Problem for Quasiadditive Set Funtions and Radon-Nikodym Derivatives*, Transactions of the American Mathematical Society, Vol. 102, pp. 114-146, 1962.

[34] Cinquini, S., *Sopra i Fondamenti di una Classe di Problemi Variazionali dello Spazio*, Rendiconti del Circolo Matematico di Palermo, Vol. 6, pp. 271-288, 1957.

[35] Cinquini, S., *Sopra l'Esistenza dell'Estremo per una Classe di Integrali Curvilinei in forma Parametrica*, Annali di Matematica Pura ed Applicata, Vol. 49, pp. 25-72, 1960.

[36] Cinquini, S., *Sopra la Semicontinuità di una Classe di Integrali del Calcolo delle Variazioni, I e II*, Bollettino dell'Unione Matematica Italiana, Vol. 11, pp. 574-588, 1975, and Vol. 12, pp. 106-116, 1975.

[37] Faedo, S., *Il Calcolo delle Variazioni per gli Integrali di Fubini-Tonelli*, Atti del Convegno Lagrangiano dell'Accademia di Scienze di Torino, Torino, Italy, pp. 61-82, 1964.

[38] Gori, C., *La Variazione Generalizzata con Peso per Applicazioni a Valori in Spazi di Banach*, Rendiconti del Circolo Matematico di Palermo, Vol. 26, pp. 275-288, 1977.

[39] Kaiser, P.J., *Length and Variation with respect to a Measure*, Atti del Seminario Matematico e Fisico dell'Università di Modena, Vol. 24, pp. 221-235, 1975.

[40] McShane, E.J., *Existence Theorems for Ordinary Problems of the Calculus of Variations*, Annali della Scuola Normale Superiore di Pisa, Vol. 3, pp. 181-211, 1934.

[41] Menger, K., *Calculus of Variations dans les Espaces Distancés Genéraux*, C.R. Paris, 202, 1936, 1007.

[42] Pucci, P., *Integrali di Riemann e di Burkill-Cesari*, Rendiconti di Matematica, Vol. 3, pp. 253-275, 1983.

[43] Radó, T., *Length and Area*, American Mathematical Society, Colloquium Publications, Providence, Rhode Island, Vol. 30, 1948.

[44] Ragni, M., *Sull'Approssimazione e Rappresentazione dell'Integrale di Burkill Classico e Generalizzato*, Atti del Seminario Matematico e Fisico dell'Università di Modena, Vol. 28, pp. 112-125, 1979.

[45] Serrin, J., *On the Definition and Properties of Certain Variational Integrals*, Transactions of the American Mathematical Society, Vol. 101, pp. 139-267, 1961.

[46] Tonelli, L., *Sui Minimi e Massimi Assoluti del Calcolo delle Variazioni*, Rendiconti del Circolo Matematico di Palermo, Vol. 32, pp. 297-337, 1911.

[47] Tonelli, L., *Sulle Funzioni di Intervallo*, Annali della Scuola Normale Superiore di Pisa, Vol. 8, pp. 1-13, 1939.

[48] Vinti, C., *L'integrale di Weierstrass*, Istituto Lombardo, Accademia di Scienze e Lettere, Vol. 92, pp. 423-434, 1958.

[49] Vinti, C., *L'integrale di Weierstrass e l'Integrale del Calcolo delle Variazioni in Forma Ordinaria*, Atti dell'Accademia di Scienze, Lettere, ed Arti di Palermo, Vol. 19, pp. 51-62, 1958-59.

[50] Vinti, C., *Espressioni che danno l'Area di una Superficie* $z = f(x,y)$ *in Relazione al Passaggio al Limite sotto il Segno d'Integrale*, Annali della Scuola Normale Superiore di Pisa, Vol. 14, pp. 103-132, 1960.

[51] Vinti, C., *L'integrale di Fubini-Tonelli nel Senso di Weierstrass, I, Caso Parametrico*, Annali della Scuola Normale Superiore di Pisa, Vol. 22, pp. 229-263, 1968.

[52] Vinti, C., *L'Integrale di Fubini-Tonelli nel senso di Weierstrass, II, Caso Ordinario*, Annali della Scuola Normale Superiore di Pisa, Vol. 22, pp. 335-376, 1968.

[53] Vinti, C., *Espressioni che danno l'Area di una Superficie*, Atti del Seminario Matematico e Fisico dell'Università di Modena, Vol. 17, pp. 289-350, 1968.

[54] Vinti, C., *L'Integrale di Weierstrass-Burkill*, Atti del Seminario Matematico e Fisico dell'Università di Modena, Vol. 18, pp. 295-316, 1969.

[55] Vinti, C., *Teoremi di esistenza, rappresentazione e approssimazione per l'integrale del Calcolo delle Variazioni*, Atti del Convegno celebrativo dell'80° anniversario della nascita di R. Calapso,

Taormina 1-4.4. 1981.

[56] Vinti, C., *Nonlinear integration and Weierstrass integral over a manifold: connections with theorems on Martingales*, J.O.T.A., Vol. 41, pp. 213-237, 1983.

[57] Warner, G., *The Generalized Weierstrass-Type Integral* $\int f(\xi,\phi)$, Annali della Scuola Normale Superiore di Pisa, Vol. 22, pp. 163-192, 1968.

[58] Warner, G., *The Burkill-Cesari Integral*, Duke Mathematical Journal, Vol. 35, pp. 61-78, 1968.

[59] Young, L.C., *An Expression Connected with the Area of a Surface* z = f(x,y), Duke Mathematical Journal, Vol. 11, pp. 43-57, 1944.

C. Vinti
Departimento di Matematica
Università degli Studi
06100
Perugia
Italy